Handbook of Laser-Based Sustainable Surface Modification and Manufacturing Techniques

This handbook provides an insight into the advancements in surface engineering methods, addressing the microstructural features, properties, mechanisms of surface degradation failures, and tribological performance of the components. Emphasis is placed on the use of laser cladding methods because they make it simple to deposit new classes of materials such nanocomposites, nanotubes, and smart materials.

Handbook of Laser-Based Sustainable Surface Modification and Manufacturing Techniques discusses the main mechanism behind the surface degradation of structural components in strenuous environments. It highlights the capacity of laser cladding to operate on a wide range of substrate materials and shapes as well as presents how laser cladding can offer new possibilities in the reconditioning of components and how, in many cases, these approaches are the only solution for economic efficiency. This handbook illustrates how the type of laser, laser optics, and the parameters of the process can be efficiently selected, and thus the power of laser cladding and its applications can be increased. The standard methods of testing used for various types of biomedical devices and tools, as well as the advantages of combining laser cladding with simultaneous induction heating, are also described in this handbook.

Features:

- Discusses the role of claddings fabricated with laser technique to withstand wear and corrosion
- Highlights the role of laser in the manufacturing of alloys and recent advancements in laser based additive manufacturing processes
- Presents the possibilities, applications, and challenges in laser surfacing
- Illustrates the post-treatments of powders and coatings and case studies related to laser surface technology
- Offers the standard methods of testing applied to various types of biomedical devices and tools
- Goes over the advantages of combining laser cladding with simultaneous induction heating

The technical outcomes of these surface engineering methods are helpful for academics, students, and professionals who are working in this field, as this enlightens their understanding of the performance of these latest processes. The audience is broad and multidisciplinary.

Sustainable Manufacturing Technologies: Additive, Subtractive, and Hybrid

Series Editor Chander Prakash, Sunpreet Singh, Seeram Ramakrishna, and Linda Yongling Wu

This book series offers the reader comprehensive insights of recent research breakthroughs in additive, subtractive, and hybrid technologies while emphasizing their sustainability aspects. Sustainability has become an integral part of all manufacturing enterprises to provide various techno-social pathways toward developing environmentally friendly manufacturing practices. It has also been found that numerous manufacturing firms are still reluctant to upgrade their conventional practices to sophisticated sustainable approaches. Therefore, this new book series is aimed to provide a globalized platform to share innovative manufacturing mythologies and technologies. The books will encourage the eminent issues of the conventional and non-conventional manufacturing technologies and cover recent innovations.

Advances in Manufacturing Technology: Computational Materials Processing and Characterization
Edited by Rupinder Singh, Sukhdeep Singh Dhami, and B. S. Pabla

Additive Manufacturing for Plastic Recycling: Efforts in Boosting a Circular Economy
Edited by Rupinder Singh and Ranvijay Kumar

Additive Manufacturing Processes in Biomedical Engineering: Advanced Fabrication Methods and Rapid Tooling Techniques
Edited by Atul Babbar, Ankit Sharma, Vivek Jain, and Dheeraj Gupta

Additive Manufacturing of Polymers for Tissue Engineering: Fundamentals, Applications, and Future Advancements
Edited by Atul Babbar, Ranvijay Kumar, Vikas Dhawan, Nishant Ranjan, and Ankit Sharma

Sustainable Advanced Manufacturing and Materials Processing: Methods and Technologies
Edited by Sarbjeet Kaushal, Ishbir Singh, Satnam Singh, and Ankit Gupta

3D Printing of Sensors, Actuators, and Antennas for Low-Cost Product Manufacturing
Edited by Rupinder Singh, Balwinder Singh Dhaliwal, and Shyam Sundar Pattnaik

For more information on this series, please visit: https://www.routledge.com/Sustainable-Manufacturing-Technologies-Additive-Subtractive-and-Hybrid/book-series/CRCSMTASH

Handbook of Laser-Based Sustainable Surface Modification and Manufacturing Techniques

Edited by
Hitesh Vasudev
Chander Prakash

CRC Press
Taylor & Francis Group
Boca Raton London New York

CRC Press is an imprint of the
Taylor & Francis Group, an **informa** business

First edition published 2023
by CRC Press
6000 Broken Sound Parkway NW, Suite 300, Boca Raton, FL 33487-2742

and by CRC Press
4 Park Square, Milton Park, Abingdon, Oxon, OX14 4RN

CRC Press is an imprint of Taylor & Francis Group, LLC

© 2023 selection and editorial matter, Hitesh Vasudev and Chander Prakash; individual chapters, the contributors

ISBN: 9781032387673 (hbk)
ISBN: 9781032389097 (pbk)
ISBN: 9781003347408 (ebk)

DOI: 10.1201/9781003347408

Typeset in Times
by codeMantra

Contents

Preface

HANDBOOK OF LASER-BASED SUSTAINABLE SURFACE MODIFICATION

Laser technologies play a role in sustainable manufacturing processes. As society's already-staggering demand for materials and goods continues to grow, the companies that produce these materials face increasing pressure to protect the global environment during manufacturing processes. Laser scientists and other photonics researchers certainly recognize the growing urgency to reduce costs and wastes while increasing efficiency and quality. In recent years with up-gradation of technology, demand for the enhanced efficiency and performance of the structural component also increases. In this context, a great deal of research and development activities have been conducted to find out the ways in order to enhance the mechanical and tribological properties of the components' surface during the performance. The selection of a material for the strenuous conditions requires specific solutions – a material that has a combination of mechanical, physical, and chemical properties is considered to be the most efficient. Since most of the industrial applications commonly deal with surface of the structural parts, applying a suitable surface treatment improves the desired properties to the level required and can minimize the surface degradation-related failures. Therefore, it is important to understand the nature of all types of environmental degradation of metals and alloys to take the preventive measures against metal loss and failures to ensure safety and reliability. Currently, many technologies and characterization techniques are being developed to improve the performance of surface to next level.

This book aims to provide an insight into the research advancements in surface engineering methods, addressing mainly the microstructural features, properties, mechanisms of surface degradation failures, and tribological performance of the components. Greater emphasis is placed on laser cladding methods as by using laser cladding a new class of materials like nano-composites, nano-tubes, and smart materials can be easily deposited.

The technical outcomes of these surface engineering methods will be helpful for the academics, students, and professionals who are working in this field as this enlightens their understanding about the performance of these latest processes. Hence, audience is broad with multidisciplinary approach.

Editors

Prof. Hitesh Vasudev, is currently working as professor in the School of Mechanical Engineering and Division of Research and Development at Lovely Professional University, Phagwara, India. He received a Ph.D. in Mechanical Engineering from Guru Nanak Dev Engineering College (affiliated to IKGPTU-Jalandhar), Ludhiana, India in 2018. His area of research is thermal spray coatings, especially for the development of new materials used for high-temperature erosion and oxidation resistance and microwave processing of materials.

He has contributed extensively in thermal spray coatings in repute journals including *Surface Coatings and Technology, Materials Today Communications, Engineering Failure Analysis, Journal of Cleaner Production, Surface Topography: Metrology and Properties, Journal of Failure Prevention and Control* and *International Journal of Surface Engineering and Interdisciplinary Materials Science* under the flagship of various publication groups such as Elsevier, Taylor & Francis Group, Springer nature, IGI Global, and InTech Open. Moreover, he is a dedicated reviewer of reputed journals such as *Surface Coatings and Technology, Ceramics International, Journal of Material Engineering Performance, Engineering Failure Analysis, Surface Topography: Metrology and Properties, Material Research Express, Engineering Research Express* and *IGI Global*. He has authored more than 30 international publications in various international journals and conferences.

He has published 15 chapters in various books related to surface engineering and manufacturing processes. He has also published a unique patent in the field of thermal spraying. He has teaching experience of more than 8 years. He received a "Research Excellence Award" in 2019, 2020, and 2021 at Lovely Professional University, Phagwara, India. He has organized a national conference and has participated in many international conferences.

Dr. Chander Prakash is a professor and dean in the Division of Research & Development/School of Mechanical Engineering at Lovely Professional University, India. He graduated from the Panjab University Chandigarh in 2016 with Ph.D. in Mechanical Engineering. He is also an Adjunct Professor (Honorary position) at the Institute for Computational Science, Ton Duc Thang University, Vietnam.

He is a dedicated teacher who embraces student-centric approaches, providing experiential learning to his students. He is a passionate researcher with diversified research interests – developing materials for biomedical and healthcare applications, additive manufacturing, and developing and exploring new cost-effective manufacturing technologies for biomedical industries. To date, he has published more than 325 scientific articles in various peer-reviewed reputed top-notch journals, conferences, and books in Materials Science and Manufacturing.

Dr. Prakash is a highly cited researcher at the international level, and he has 5513 citations, an H-index of 41, and an i-10 index of 123. He is one of the Top 1% of leading scientists in Mechanical and Aerospace Engineering in India, as per Research. com. He holds the 38th rank in India and the 1590th rank in the world. He also consistently appeared in the top 2% of researchers as per Stanford Study in 2021 and 2022.

Dr. Prakash edited/authored 25 books, serving as a series editor of 2 books, serving as guest editor, and serving as an editorial board member of journals. He is working on research commercialization and has published 18 patents. His four patents were granted. Dr. Prakash raised a fund of Rs. 21.50 Lakhs for "Wire Arc Additive Manufacturing for Industrial Application" under the PRISM scheme of the Ministry of Science & Technology, India in 2022. He received a grant of Rs. 3.00 Lakhs from UKIERI-DST for Partnership Development Workshops on "Medical Additive Manufacturing Cost-effective and Sustainable Solutions for Innovative Development of High-Value Added Products and Services Testing" in collaboration with University of Greenwich, UK and Cardiff University, UK.

Dr. Prakash has organized three series of the signature International Conference on Functional Materials Manufacturing and Performances (ICFMMP) in 2019, 2021, and 2022. Recognizing his contribution to research and development, Lovely Professional University awarded him Research Excellence Award for the best and most highly productive researcher in 2019, 2020, and 2021.

Contributors

Natarajan Jeyaprakash
School of Mechanical and Electrical
  Engineering
China University of Mining and
  Technology
Xuzhou City, China

Sundara Subramanian Karuppasamy
Graduate Institute of Manufacturing
  Technology
National Taipei University of
  Technology
Taipei, Taiwan

Amrinder Mehta
Division of Research and Development
Lovely Professional University
Phagwara, India

Chander Prakash
Division of Research and Development
Lovely Professional University
Phagwara, India

Gaurav Prashar
Department of Mechanical Engineering
Rayat Bahra Institute of Engineering
  and Nano Technology
Hoshiarpur, India

Kuldeep K. Saxena
Division of Research and Development
Lovely Professional University
Phagwara, India

Jashanpreet Singh
University Center for Research and
  Development
Chandigarh University
Mohali, India

Sharanjit Singh
DAV University
Jalandhar, India

Hitesh Vasudev
Division of Research and Development
School of Mechanical Engineering
Lovely Professional University
Punjab, India

Che-Hua Yang
Graduate Institute of Manufacturing
  Technology
National Taipei University of
  Technology
Taipei, Taiwan

1 Application of Wear-Resistant Laser Claddings

Natarajan Jeyaprakash
China University of Mining and Technology

Sundara Subramanian Karuppasamy and Che-Hua Yang
National Taipei University of Technology

CONTENTS

DOI: 10.1201/9781003347408-1

1

1.1   INTRODUCTION

Metal alloys are formed by combining one base metal with one or more metal/non-metals in such a way that the formed alloy experiences better performance compared to the base metal. In general, various types of alloys have been implemented in fabricating the components that are involved in automobiles, aerospace, marine, and nuclear sectors. Each alloy has discrete characteristics that are well adapted to the prevailing environmental conditions and provide excellent performance owing to its characteristics [1,2]. Most of the sectors got benefitted by means of Fe-, Ni-, Al-, Ti-, and Co-based alloys because of their predominant features that are not exhibited by other alloys. Iron alloys have better magnetic properties, higher strength, greater toughness, etc. and are used in abundant applications [3]. Nickel alloys are well known for their heat-resistant behaviour and majorly employed in high-temperature environments like nuclear reactors, steam turbines, power plants, etc. [4]. Aluminium alloys are predominantly used in packaging and automobile sectors due to their anti-corrosion behaviour, high-end processability, and greater specific strength, whereas titanium alloys are implemented in the medical field because of their biocompatible features [5,6]. Thus, the above-mentioned alloys dominated almost all sectors for many years.

Besides their characteristics, the components/parts made from these alloys encounter major issues after prolonged usage in the working environment. These issues were due to the physical or chemical or mechanical phenomena which will pave the way for the catastrophic failure of these components followed by property loss. And in some cases, these issues will lead to fatal accidents that could be dangerous for mankind. Among the issues, the most critical issue is wear. Wear can be defined as the process of material removal or deformation of the surface by means of contacting with other surface in a mechanical manner. This phenomenon is unavoidable and caused by mechanical or chemical constraints between the surfaces [7,8]. Based on the relative motion between the surfaces, damage severity, interaction, and damage mechanism, the wear can be classified into different categories such as abrasive, adhesive, erosive, corrosive, fatigue, and fretting wear [9,10]. In addition, when two materials are in contact, there will be formation of debris-like substances due to the wear between them and termed as 'wear debris'. This debris also acts as a key factor to understand the mode of wear mechanism during the wear process [11].

Since these types of wear phenomena and generation of wear debris are inevitable, much funds are allocated to overcome the wear in components. Surface engineering plays a vital role in altering the damaged surfaces in an economic manner. It involves many surface modification techniques that can repair the worn-out surfaces effectively. This technique minimizes the need for replacement of the worn-out parts, thereby coating it with several coating materials that provide significant resistance towards wear. The major techniques that are implemented to repair the worn-out surfaces are mentioned as follows: physical vapour deposition, chemical vapour deposition, micro-arc oxidation, electrodeposition, sol-gel coating, thermal spraying, and laser cladding [12,13]. Each technique has its own pros and cons, but laser cladding is more advantageous than the other techniques. It has better metallurgical bonding between the coating layer and the base material, minimal heat-affected zone, less distortion and dilution effect, better uniformity in coating, cost-effective, etc. [14]. This process involves a high-energy laser beam that irradiates the surface of the base material, thereby generating a molten melt pool with minimal heat-affected zone. The coating material is introduced in this melt pool that gets uniformly deposited on the base material. After solidification, a fine coating layer was achieved, and depending on the coating material, the layer will exhibit better characteristics than the base material [15]. Figure 1.1 represents the schematic illustration of the laser cladding process.

The coating/cladding material decides the performance of the cladded layer formed on the base materials; adequate attention is said to be given while preferring the suitable cladding material. In most cases, superalloys are preferred as cladding materials because of their remarkable characteristics. These alloys are often termed as 'high-performance alloys' with excellent strength, greater resistance towards wear, heat, corrosion, and creep deformation. It can deliver efficient performance even near to its melting point and most often used in applications where heat-resistant behaviour is the major criterion. Hence, these superalloys are used in fabricating the components of aero turbines, surgical implants, gas turbines, power plants, gasification and liquefaction systems, etc. [16,17]. These alloys can be of different types. Mostly used are nickel-based superalloys (Inconel, Colmonoy, Nimonic, and

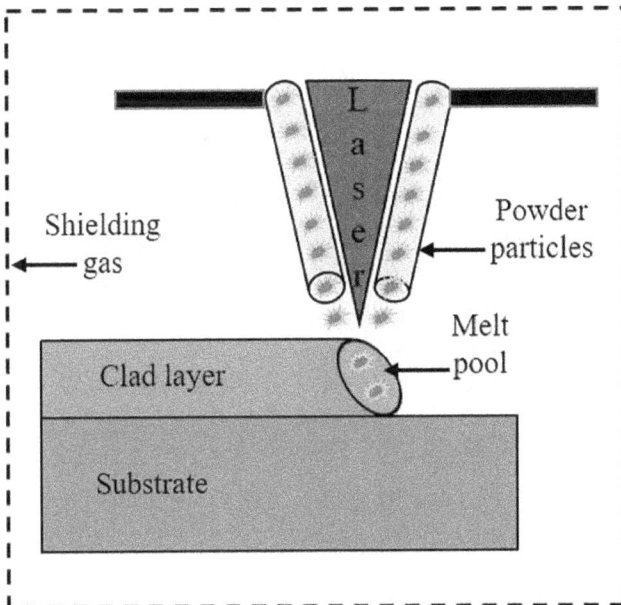

FIGURE 1.1   Laser cladding process.

Hastelloy series), cobalt-based superalloys (Stellite series), and iron-based superalloys [18]. Other than these three categories, other alloys are also used as cladding
materials depending on the requirements.

In recent decades, many researchers have evaluated the performance of the cladded layer made by using the above-mentioned superalloys on different substrates
(base material) and compared the tensile, wear, hardness, fatigue, and other tribological features. Weng et al. [19] analysed the cladding behaviour of various coating
materials on titanium alloys and reported that the secondary precipitates formed
after solidification are responsible for the improvement of wear-resistant characteristics. Liu et al. [20] and Das [21] studied the research aspects of laser cladding on
magnesium alloys. Their study concluded that by means of laser cladding, the tribological properties were improved at a maximum level than the other methods.
Singh et al. [22] and More et al. [23] examined the effectiveness of laser cladding
technique for applications involving erosive wear phenomenon and reported that this
technique provides better anti-erosive wear characteristics and extends the durability of the components. Arif et al. [24] interrogated the research status, recent trends,
and applications of cladded layer made from high-entropy alloys and its mechanical behaviours. In addition, laser cladding technique is reported to minimize the
wear in the near-surface regions of self-lubricating composites [25]. Hence, from the
above literatures, it can be concluded that the laser cladding serves as the efficient
and effective technique for improving the wear-resistant behaviours of the substrate
materials. Compared to the other surface treatment techniques, laser cladding is suitable for enhancing the anti-wear characteristics and aids in extending the working
life of the components.

This chapter is focused on investigating the wear-resistant characteristics, wear
mechanisms, and hardness of the cladded layer formed by various coating materials on different substrates. The Scanning Electron Microscope (SEM) images of
the cladded powder used for coating on the different substrates were provided. The
microstructure at the cladded layer formed on various substrates was examined by
means of Field Emission Scanning Electron Microscope (FESEM) images captured
at the clad layer. With the aid of pin on disk apparatus, the wear test was performed
on the different cladded layers to examine the wear-resistant characteristics exhibited by those layers. The wear mechanisms behind the wear test were schematically
represented and explained with the help of worn-out SEM images captured after the
wear test.

1.2 WEAR TYPES

This section gives a brief introduction about the wear types.

1.2.1 Abrasive Wear

Abrasive wear happens when the hard surface comes in contact with the soft surface.
This type of wear is common, and in developed nations, this wear consumes around
4% in the gross national production of components. The type of contact between

the two surfaces decides the mode of abrasive wear that takes place. This abrasive wear phenomenon consists of two types of modes: two-body and three-body abrasive wears. Two-body abrasive wear takes place when the hard surface slides on the soft surface. The effect of wear debris (abrasive particles) that are entrapped within these surfaces results in three-body abrasive wear [26]. The major reasons for this type of wear are heavy-load conditions, uneven rotating speed, incorrect heat profile, enormous back pressure, and so on. One of the most common examples of abrasive wear is the wear experienced by gear teeth.

1.2.2 Adhesive Wear

Adhesive wear takes place during the frictional contact between the two surfaces, and the material removal is observed on one or both surfaces, and this wear debris is resulted from the strong adhesive force that exists between these surfaces [27]. Depending on the damage severity level, the adhesive wear is classified into normal, moderate, and excessive adhesive wears. Inadequate loading, high temperature, and more moisture in the wear zone serve as the root causes of this type of wear. Parts like bearings, engine pistons, power transmission systems (gearbox), and piston cylinders were affected by this type of wear.

1.2.3 Fatigue Wear

Fatigue wear is defined as the process of removal from the surface by means of strains that are induced for a certain limit of time or cycles. Cracks are initially formed and further penetrate, which results in fatigue on the surface. This type of wear is caused by the sliding or rolling contact between the surfaces [28].

1.2.4 Fretting Wear

Fretting wear is considered as the special type of sliding wear phenomenon where the sliding motion between the two surfaces is said to be in a lower amplitude range (mostly 1–150 µm). This type of wear arises from the elastic deformation or vibration between the surfaces, and it can initiate the nucleation zone for cracks formation [29]. The wear debris formed may have a sticking effect rather than material removal, and this effect affects the motion. Usually, fretting wear is experienced by the surfaces/parts that have various values of coefficient of thermal expansion like driveshafts, cylinder head gaskets, etc.

1.2.5 Erosive and Corrosive Wears

Erosive wear can be described as the process in which the material is removed from the surface with the aid of external particles [30]. These particles are considered as the erosive particles that cause wear while moving with adequate velocity. The particle size, shape, and flux rate are considered to be the major factors that influence the erosive wear. This type of wear can be seen in turbine blades of aircraft and ships, material removal in the pipes, etc. On the other hand, corrosive wear happens due to the electrochemical reaction that takes place between the surface and corrosive media [30]. During sliding, the oxide film formed on the surface initiates to break and acts as wear debris between the surfaces. Corrosive wear can be noticed in the rotating parts of marine equipment.

1.3   WEAR ISSUES IN DIFFERENT SECTORS

Almost all the sectors encounter wear problems. Some of the critical wear issues that belong to different sectors are listed as follows. In nuclear power plants, major wear issues occur in pressurized water reactor core components like control assembly and control drive mechanisms [31]. Mostly, the control assembly consists of stainless steel tubes with guides. Greater water flow rate through these tubes will influence the guides, which will pave the way for wear. More amount of indent marks (wear scares) have been noticed and need to be fixed in order to avoid the disaster. Also, the control drive mechanism contains the drive rod and gripper latch arms. During interaction, wear phenomenon is observed in the area where the latch and rod interact mechanically [32]. Thus, in both control assembly and control drive mechanisms, maximum wear rates are noticed, which will aid for the failures. The wear debris that arises from this wear process is carried by the high-velocity water to different parts in the reactor and leads to many problems that could be danger to mankind. Moreover, fatigue and fretting type of wear were reported due to the improper loading condition in the tubes used for steam generators and fuel transmission [33].

Various types of wear phenomena were observed in the automobile sector. The sliding contact exists in the parts of power generation systems in automobile like engine piston, piston shafts, piston rings, and valves. Adhesive and abrasive wears have been noticed in these parts because of the prevailing conditions like high heat, inadequate rotation of shafts, cold starts, fuel impurities, etc. [34]. The gears present in the power transmission system encounter adhesive wear due to inadequate loading conditions. Also, the automotive braking system contains parts made by using high-friction materials. The wear particulates that were generated while applying the brakes in automobiles were released into the environment. This wear debris constitutes for 35% of the brake pad loss according to the reports and causes air-borne problems in humans [35].

Erosive wear has been reported in the propeller blades of aircraft and ships. The minute particulates present in the air interact with the propeller blades of aircraft engines at high velocity. This velocity is enough to damage the blades, resulting in erosive wear in blades. On the other hand, the water particulates come in contact with the propeller blade of ships at high flow rate, and this impact is referred as 'erosive wear' [36]. Corrosive wear dominates the pipeline industry for many years. The transporting fluids (corrosive media) consist of corrosive particulates (slurry) like chlorides, salt deposits, and sulphates, which will react with the inner surface of the pipelines. These particulates will induce localized corrosive wear attacks in pipes, thereby degrading them to a maximum extent [37]. In rails, the rolling fatigue wear is observed due to the rolling contact between the rails and wheel. The rails have a maximum wear rate under excessive loads, high-speed and frequent braking that may lead to the nucleation of cracks on the rails [38]. The biomedical industry is also reported to be affected by the wear phenomena. Most of the biomedical implants used as heart valves, bone repair plates, pacemakers, and tooth experience wear after long-term usage [39]. Thus, most of the sectors have wear issues and need to be fixed before failure occurs.

1.4 WEAR-RESISTANT CLADDING POWDERS

This section deals with the cladding powders used for coating on different substrates. The size and shape distribution of the powders was examined with the aid of FESEM images. In addition, the characteristics of each powder along with their uses have been addressed.

1.4.1 INCONEL 625 POWDER

Inconel 625 alloy belongs to the family of nickel-based superalloys formed by solid solution strengthening and precipitation hardening process. This alloy has the face-centred cubic lattice of γ-nickel phase. This alloy is a combination of heavy and light refractory metals like tantalum, niobium, and molybdenum. These elements combine with other elements like chromium, iron, carbon, silicon, phosphorous, aluminium, and manganese to form the Inconel 625 alloy. This combination of elements made this alloy to resist heat even near its melting point (1290°C–1350°C). In spite of its heat-resistant behaviour, this alloy has other features like high strength, better tensile properties, excellent fabricability, greater wear, and corrosion resistance [40]. These remarkable characteristics made this alloy to get implemented in the production of parts that are used in nuclear reactors, aircraft, marine equipment, turbines, and so on. Figure 1.2 shows the FESEM image of the Inconel 625 powder used for coating on the substrate material with an average size of $116.5 \pm 17\,\mu$m. The elemental composition of the Inconel 625 is listed in Table 1.1.

1.4.2 COLMONOY 6 POWDER

Another type of nickel-based superalloy that is well known for its abrasive wear resistance behaviour is the Colmonoy 6 alloy. It has higher percentages of chromium along with boron, carbon, iron, and silicon. This alloy resists wear to a maximum extent due to the secondary phases that are formed after solidification. This NiCrSiB alloy has excellent mechanical features like greater toughness, better fatigue and creep resistance, higher melting point (1030°C), better yield strength, etc. This alloy is used in fabricating heat exchangers, extruders, coal-fired boilers, engine shafts, piston, etc. [41]. Figure 1.3 represents the FESEM image of Colmonoy 6 powder particles that are used in the laser cladding process. The average size of those Colmonoy 6 particles is measured to be $110 \pm 40\,\mu$m, and the chemical composition of the Colmonoy 6 alloy is mentioned in Table 1.1.

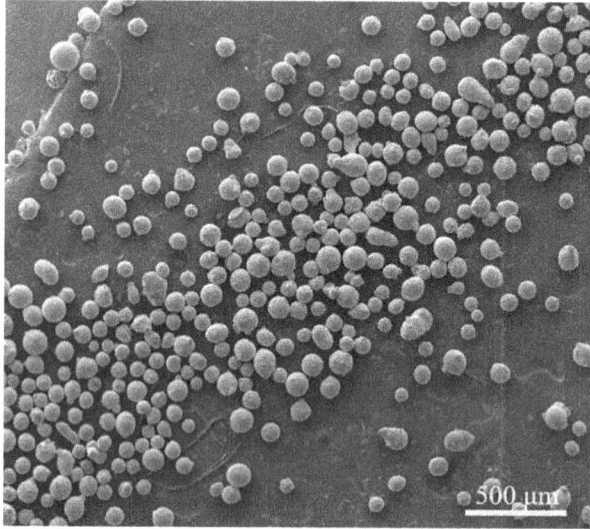

FIGURE 1.2   FESEM image of Inconel 625 powder.

TABLE 1.1
Elemental Composition of Powder Materials

Material	Ni (%)	Co (%)	Cr (%)	Fe (%)	B (%)	Si (%)	C (%)	Mn (%)	Mo (%)	W (%)
Inconel 625	BAL	1	23	5	–	0.05	0.10	0.50	10	–
Colmonoy 6	BAL	–	14.3	4	3	4.25	0.70	–	–	–
Colmonoy 5	BAL	–	13.8	4.8	2.1	3.3	0.45	–	–	–
Stellite 6	0.62	BAL	30	0.82	–	1.27	1.22	0.14	–	4.39
H13 steel	0.3	–	5	BAL	–	1.2	0.45	0.5	1.75	–
SS 420	–	–	13.5	BAL	–	0.9	0.15	1	0.5	–

1.4.3   COLMONOY 5 POWDER

Colmonoy 5 alloy is a type of superalloy that comes under the Colmonoy series of nickel-based superalloys. This alloy differs from the Colmonoy 6 alloy in the elemental composition of elements like nickel, chromium, boron, carbon, iron, and silicon. This alloy is reported to have the same melting point as in Colmonoy 6. After cladding, the microstructure of the laser-cladded Colmonoy 5 layer is rich in chromium carbide and boride precipitates. The segregation of these precipitates in the clad layer made this alloy to exhibit excellent hardness and wear-resistant characteristics [42]. This alloy serves as the main choice where hardness and wear-resistant features are the key aspects. Thus, this alloy is used in the manufacturing of reactor core components, pressurized water reactors, furnace, steam boilers, power plants, etc. Figure 1.4 represents the FESEM image of the Colmonoy 5 powder particles

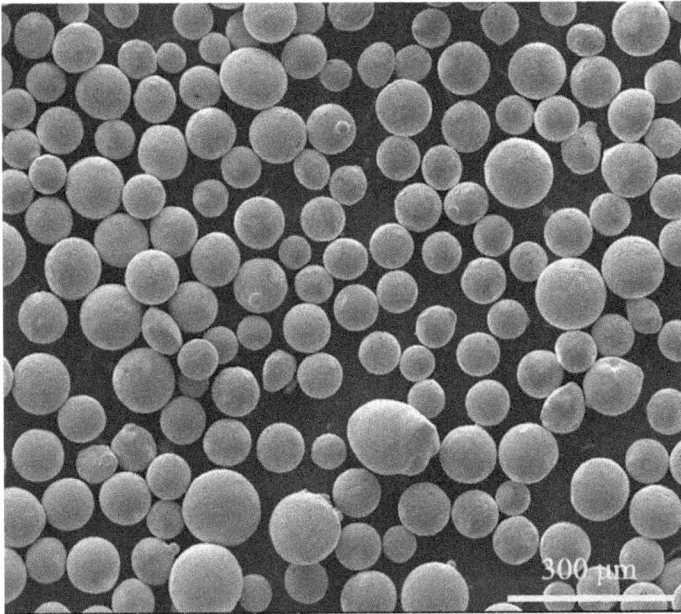

FIGURE 1.3   FESEM image of Colmonoy 6 powder.

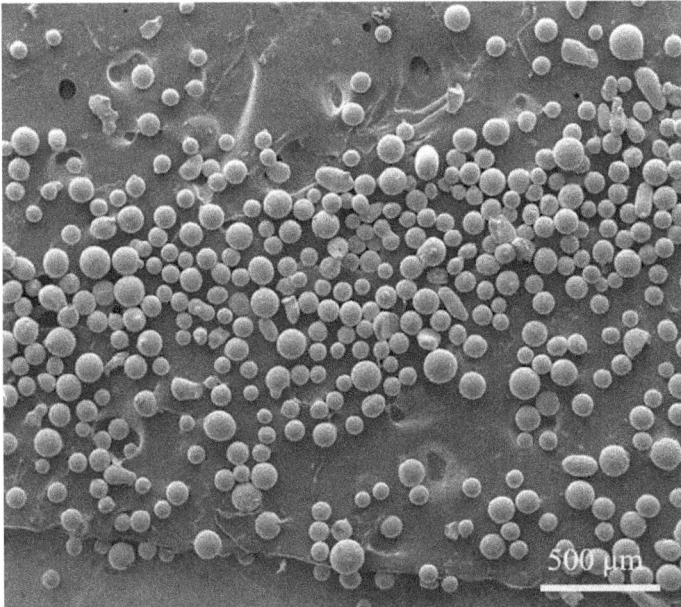

FIGURE 1.4   FESEM image of Colmonoy 5 powder.

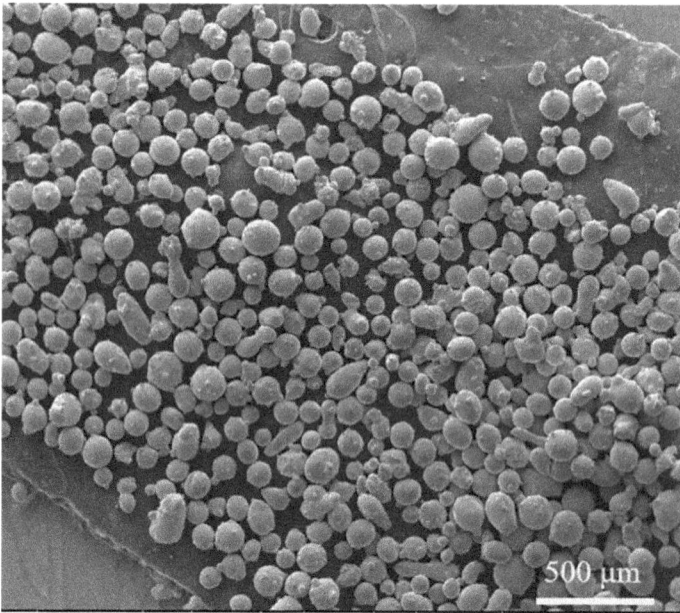

FIGURE 1.5   FESEM image of Stellite 6 powder.

(average size $= 150 \pm 25\,\mu m$), and Table 1.1 lists the chemical composition of the Colmonoy 5 alloy particles.

1.4.4   Stellite 6 Powder

Stellite 6 alloy is one among the cobalt-based superalloys with higher percentage of chromium nearly 32%. This alloy is well known for its characteristics such as higher hardness, excellent strength, higher bond strength, low porosity, better electrical resistivity and thermal conductivity, significant resistance to scratch, abrasion, cavitation erosion, and other wear types [43]. After cladding, this alloy exhibits hypo-eutectic structure that made this alloy to exhibit the above-mentioned properties. This alloy is more suitable for hardfacing process, and the coatings produced by this alloy are reported to be very effective. Hence, this alloy is used in the production of pump shafts, bearings, valve seats, rolling couples, chemically inert machine tools, impellers, and poppet valves. Figure 1.5 shows the FESEM image of the Stellite 6 powder particles with an average size of $214 \pm 25\,\mu m$, and its elemental composition is tabulated in Table 1.1.

1.4.5   H13 Steel Powder

H13 steel alloy is formed by the combination of Cr-Mo elements and is widely used as tool steel. This steel has outstanding characteristics such as high hardness, anti-thermal softening, greater toughness, better tensile properties, good stability, resistance

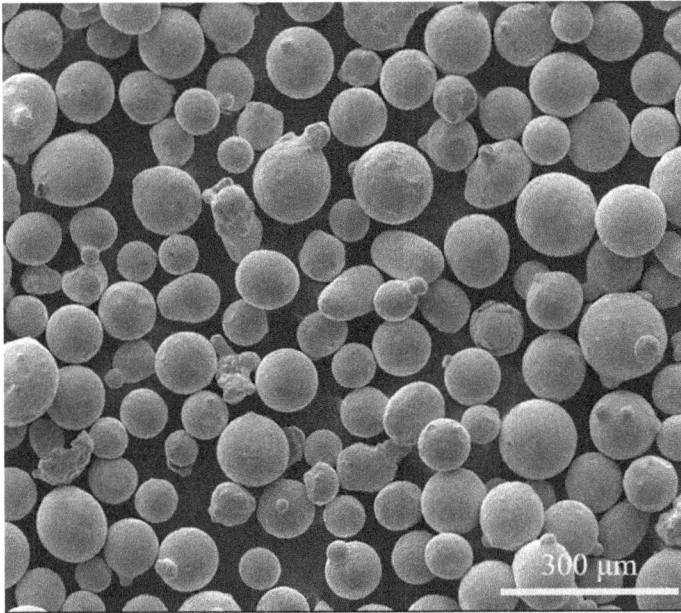

FIGURE 1.6 FESEM image of H13 steel powder.

to thermal fatigue cracks, hot corrosion cracks, and stress-induced corrosion attacks [44]. This alloy gets well suited in both hot and cold work conditions. Because of its outstanding characteristics, it is used in the fabrication of hot forging dies, sleeves of die-casting slots, dies used for extrusion, cavities of plastic moulds, sheet metal, and stamping tools. Also, this alloy is unavoidable in application that requires excellent toughness and polishability. Figure 1.6 represents the FESEM image of the H13 steel alloy powder used for the laser cladding process (average size $= 115 \pm 35\,\mu m$), and Table 1.1 lists its elemental composition. Moreover, this alloy can be heat-treated and annealed, and based on the heat-treated condition, it exhibits improved mechanical behaviours.

1.4.6 STAINLESS STEEL 420 POWDER

Stainless steel 420 belongs to the category of martensitic stainless steel. This steel is heat treatable and has superior mechanical properties. This steel resists heat up to 1510°C, which paved the way for high-temperature applications. Owing to its heat-resistant characteristics, this steel has excellent hardness, higher tensile properties, better stability, greater toughness, and resistance to fatigue, erosive, and corrosive wear [45]. This steel is implemented in the manufacturing of shear blades, needle valves, cutlery equipment, surgical instruments, etc. The FESEM image of the stainless steel 420 powder particles is shown in Figure 1.7. These powder particles have an average size of $141 \pm 20\,\mu m$, and its chemical composition is listed in Table 1.1.

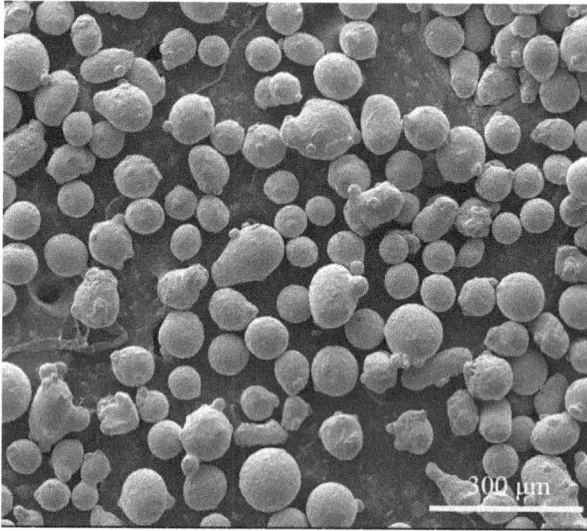

FIGURE 1.7   FESEM image of stainless steel 420 powder.

1.5   LASER CLADDING – MICROSTRUCTURE MODIFICATION

This section deals with the microstructures that are formed after solidification of different cladded layers on various substrates with the aid of FESEM images taken at the clad layer.

1.5.1   STELLITE 6 CLADDING ON INCONEL 625 SUBSTRATE

The Stellite 6 particles were introduced in the molten melt pool formed on the Inconel 625 substrate. The optimized laser cladding parameters used here are as follows: power = 1400 W; feed rate = 9 g/min; scanning speed = 600 mm/min; preheat temperature = 150°C; shielding gas flow rate = 25 L/min; and carrier gas flow = 6 SD @ 1 bar. After laser cladding, the clad specimens undergo microstructure examination with the aid of FESEM images captured at the clad layer. Figure 1.8 shows the FESEM image taken at the Stellite 6 clad layer at 10 μm magnification. From this figure, it can be seen that the microstructure of the Stellite 6 clad layer is rich in the hypoeutectic structure. Both dendritic and interdendritic phases are formed after solidification. Since Stellite 6 is a cobalt-based superalloy, the formed hypoeutectic structure is rich in α-Co matrix [46]. On further evaluation of these dendritic and interdendritic phases, it can be noticed that the dendritic phases are rich in Co, whereas the chromium-rich carbides were segregated as the interdendritic phases [47,48]. These interdendritic phases (Cr-rich carbides) aid in enhancing the tribological properties of the Stellite 6 clad layer. In particular, these carbides are responsible for wear-resistant characteristics [49]. In addition, other than these two phases, some intermetallic compounds were formed. These compounds are rich in Si, Fe, and Mn and aid in improving the mechanical behaviour of the clad layer.

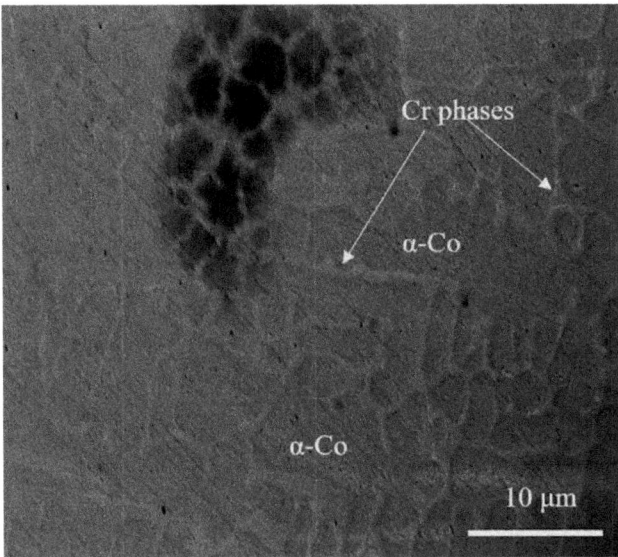

FIGURE 1.8   FESEM image of Stellite 6 cladding.

1.5.2 COLMONOY 5 CLADDING ON STAINLESS STEEL 410 SUBSTRATE

Here, the Colmonoy 5 powder particles were deposited in the melt pool of stainless steel 410 substrate. The laser parameters used to coat the Colmonoy 5 particles are as follows: power = 1000 W; feed rate = 4 g/min; scanning speed = 400 mm/min; preheat temperature = 150°C; shielding gas flow rate = 25 L/min; and carrier gas flow = 6 SD @ 1 bar. The clad samples were collected after the cladding process, and the microstructural examination was performed on the clad layer by means of the FESEM images. The captured FESEM image at the Colmonoy 5 clad layer is represented in Figure 1.9. The optimized process parameters paved the way for the formation of hard laves phase in the clad layer [50,51]. Since Colmonoy 5 alloy is a Ni-based superalloy, the formed laves phase is rich in ferric nickel and nickel borides. Along with the laves phase, the formation of dendritic and interdendritic precipitates was observed in the γ-nickel matrix [52]. These precipitates are termed as 'dark' (needle-shaped) and 'floret-shaped precipitates'. The dark precipitates are rich in chromium carbides, and floret precipitates are rich in chromium borides [53]. These chromium-rich borides are reported to enhance the anti-wear behaviour at the Colmonoy 5 clad layer. Also, other compounds like nickel silicide were formed that help in improving the hardness property. Thus, the secondary phases help in promoting the wear-resistant characteristics at the Colmonoy 5 clad layer.

1.5.3 COLMONOY 6 CLADDING ON INCONEL 625 SUBSTRATE

For improving the wear-resistant characteristics of Inconel 625 alloy, Colmonoy 6 particles were coated on the Inconel 625 substrate. The laser process parameters used

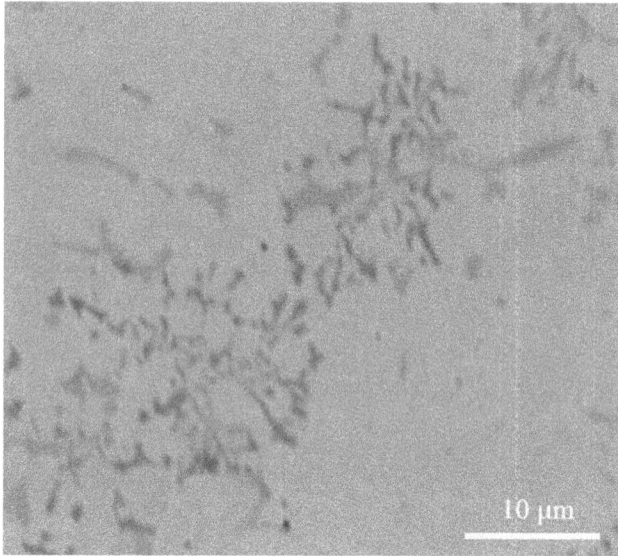

FIGURE 1.9   FESEM image of Colmonoy 5 cladding.

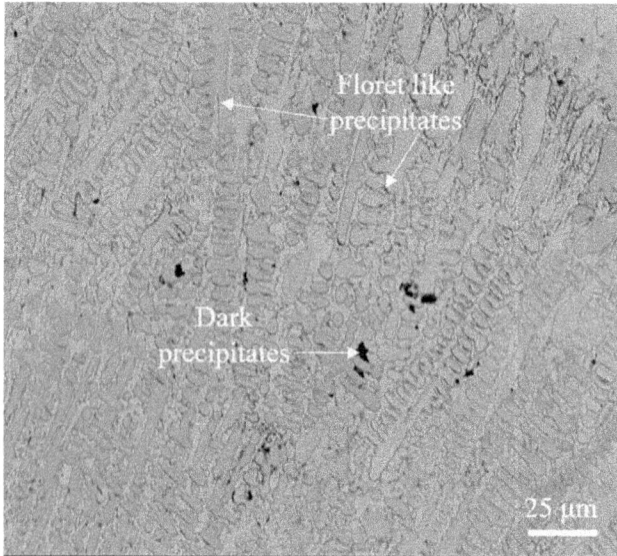

FIGURE 1.10   FESEM image of Colmonoy 6 cladding.

for this process are summarized as follows: power = 1300 W; feed rate = 6 g/min; scanning speed = 400 mm/min; preheat temperature = 100°C; shielding gas flow rate = 25 L/min; carrier gas flow = 6 SD @ 1 bar. For investigating the microstructure present at the Colmonoy 6 clad layer, the FESEM image taken at the clad layer is shown in Figure 1.10. From this figure, it can be inferred that the clad region is rich in γ-Ni solid solution [54]. This is because both the substrate and cladding material belong to the family of Ni-based superalloys. Throughout the microstructure, fine dendritic phases were formed along with the interdendritic precipitates [55]. This interdendritic precipitate consists of blocky and floret-shaped phases. Researchers reported that the blocky precipitates are rich in carbides of chromium and CrB constitutes for floret-shaped precipitates. The segregation of these secondary precipitates at the Colmonoy 6 clad layer is because of the optimized process parameters used during laser cladding. Moreover, these carbides and borides aid in facilitating the hardness and wear resistance offered by the Colmonoy 6 clad layer [56,57].

1.5.4   INCONEL 625 CLADDING ON INCONEL 625 SUBSTRATE

Some substrate materials are said to provide better wear-resistant behaviour when coated it with the same material. Considering this aspect, the Inconel 625 alloy powder is coated on the same Inconel 625 substrate. The process parameters used for coating Inconel 625 on Inconel 625 substrate are as follows: power = 1000 W; feed rate = 4 g/min; scanning speed = 400 mm/min; preheat temperature = 150°C; shielding gas flow rate = 25 L/min; carrier gas flow = 6 SD @ 1 bar. The microstructure formed on the Inconel 625 clad layer is investigated by means of optical micrograph as in Figure 1.11. From this micrograph, the clad layer consists of fine cellular microstructure [58]. These cellular grains are rich in the γ-Ni phases since both the substrate and clad material are Inconel 625 alloy. Owing to the cellular dendritic structure, the formed structure contains interdendritic precipitates. This precipitate is rich in nickel carbides, nickel silicide, and ferric nickel carbides [59,60]. These compounds were randomly segregated throughout the cellular dendritic structure.

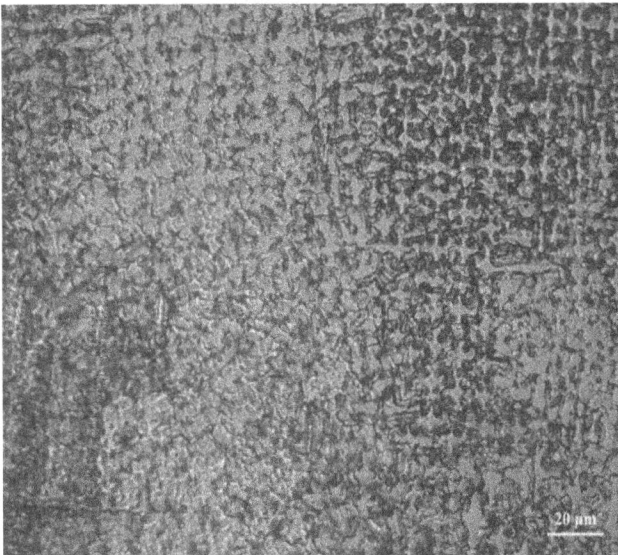

FIGURE 1.11   Optical micrograph of Inconel 625 cladding.

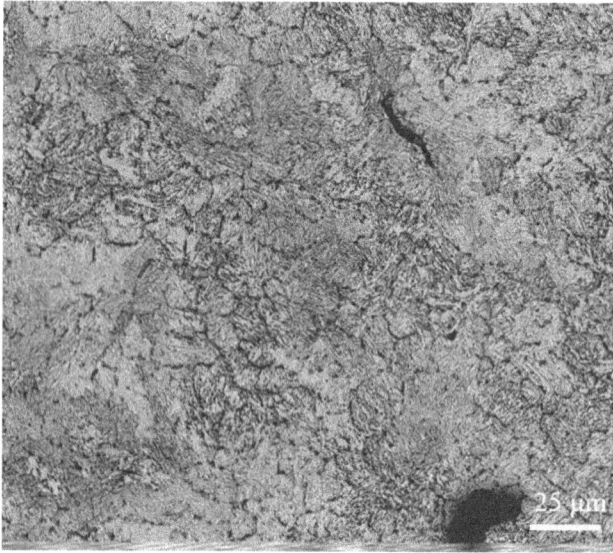

FIGURE 1.12   FESEM image of stainless steel 420 cladding.

This fine cellular dendritic morphology along with the secondary phases enhanced the anti-wear characteristics of the Inconel 625 clad layer [61].

1.5.5   Stainless Steel 420 Cladding on Stainless Steel 410 Substrate

In some cases, the substrate's tribological behaviour can be enhanced with normal alloys rather than superalloys. Stainless steel serves as a better example for this case. Here, the stainless steel 420 powder particles are cladded on the stainless steel 410 substrate material. This cladding process is carried out with the laser parameters as follows: power = 1000 W; feed rate = 4 g/min; scanning speed = 400 mm/min; preheat temperature = 150°C; shielding gas flow rate = 25 L/min; and carrier gas flow = 6 SD @ 1 bar. Figure 1.12 represents the FESEM micrograph taken at the stainless steel 420 clad layer. From this FESEM micrograph, it can be observed that the obtained microstructure consists of a closely packed needle-like structure, which resembles the traditional martensitic phase. This martensitic phase is the only existed hard phase in the cladded microstructure and offers load-carrying capacity [50,62]. Moreover, this fine distribution of needle-like acicular structure along with secondary precipitates is reported to provide better hardness and wear resistance at the clad layer [63]. These secondary precipitates contain the carbides of iron and chromium.

1.6   DRY SLIDING WEAR TEST

In this section, a dry sliding wear test was performed on the clad layers made up of different alloy powders. Here, the cladding region serves as the testing surface, and the hardened ball made up of tungsten carbide acts as the counterpart that slides on

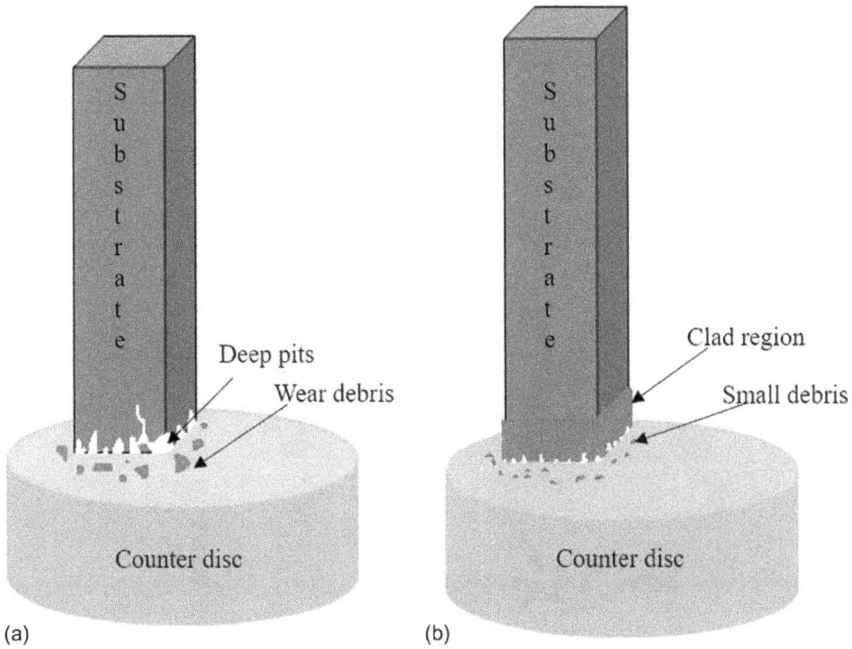

FIGURE 1.13   Wear mechanism at: (a) Substrate surface with more wear debris and (b) clad region with less wear debris.

the clad layer. Figure 1.13 shows the schematic illustration of the wear mechanism. In addition, the worn-out images captured at the clad layer were discussed for a better understanding of the wear phenomenon.

1.6.1   Stellite 6 Cladding on Inconel 625 Substrate

Here, the dry sliding wear test was carried out on the Stellite 6 cladded layer with the aid of pin on disk equipment. The wear test parameters are as follows: load = 50 N; sliding speed = 1 m/s; and sliding distance = 2000 m. After the wear test, the worn-out clad surface was captured by using FESEM microscope. Figure 1.14 shows the worn-out image of the Stellite 6 clad layer. From this figure, the wear debris, material-removed area, and wear tracks can be easily visualized. The material-removed area is indicated by the bright spots in the FESEM image. The test surface has fewer scratches, and these scratches are reported to be finer. Moreover, the wear tracks have lesser depth, which indicates that the cladded layer has resistance against wear. Both abrasive and adhesive wear phenomena have been observed in the clad layer. During the sliding contact, the abrasive particles were formed and stayed on the testing surface which initiates the abrasive wear [64]. The materials that pile up on the surface are lesser. Thus, the Stellite 6 cladded layer has better wear-resistant characteristics since the wear tracks were finer with lesser depth. This is due to the

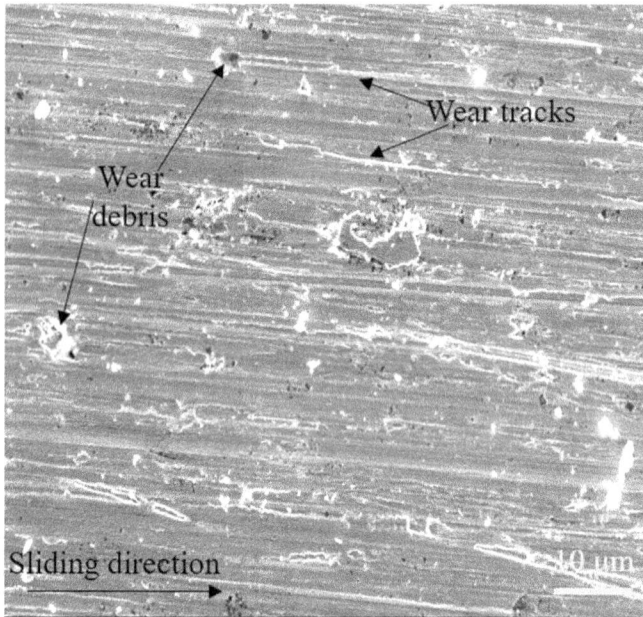

FIGURE 1.14   FESEM image of worn-out surface at the Stellite 6 clad layer.

existence of cobalt-rich dendritic and chromium-rich interdendritic (Cr-carbides) in the clad layer [65]. Also, some researchers confirmed that Stellite 6 alloy has better wear-resistant characteristics due to the combined effect of these two phases [66,67].

1.6.2   COLMONOY 5 CLADDING ON STAINLESS STEEL 410 SUBSTRATE

To evaluate the anti-wear characteristics of Colmonoy 5 alloy particles cladded on the stainless steel substrate, the dry sliding wear test was carried out at the clad layer with the parameters mentioned as follows: load = 48 N; sliding speed = 0.4 m/s; sliding distance = 2500 m. The worn-out FESEM image captured at the Colmonoy 5 clad layer is shown in Figure 1.15. It can be inferred that the worn-out surface consists of wear debris along with the wear tracks and debris-detached area [50]. The wear tracks are quite finer with low depth. Also, the debris that piles up from the Colmonoy 5 cladded surface is very low. The adhesive wear takes place during the sliding of the ball on the testing surface, and the formed debris leads to abrasive wear. Hence, both abrasive and adhesive wear might have occurred simultaneously on the test surface [68]. From the wear track depth and debris quantity, it can be confirmed that the Colmonoy 5 clad layer has significant resistance toward wear phenomena. This significant resistance against wear is due to the interdendritic (secondary) precipitates containing Cr- rich carbides and borides. Similar results have been founded by other researchers, which proved that these precipitates act as a barrier and resist wear to a greater extent [50,69,70].

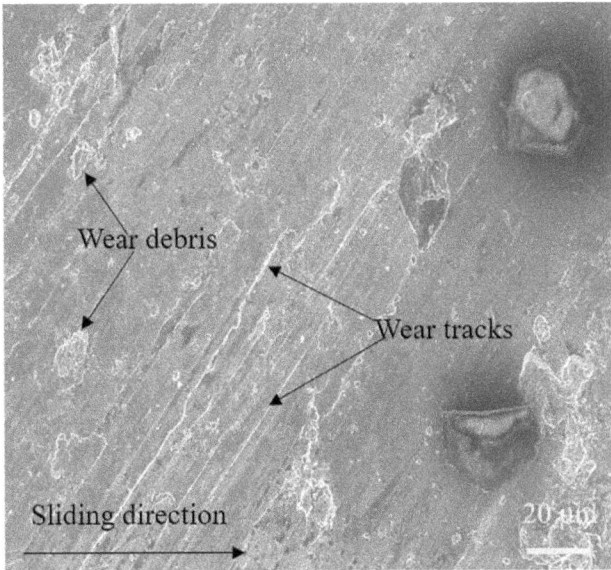

FIGURE 1.15 FESEM image of worn-out surface at the Colmonoy 5 clad layer.

1.6.3 COLMONOY 6 CLADDING ON INCONEL 625 SUBSTRATE

After cladding the Inconel 625 substrate with the Colmonoy 6 alloy powder, the wear test was performed on the cladded layer for evaluating its wear behaviour. The parameters by which the wear test was performed are as follows: load = 40 N; sliding speed = 1 m/s; and sliding distance = 2500 m. The worn-out image taken at the Colmonoy 6 clad layer is represented in Figure 1.16. During the wear test, the Colmonoy 6 clad layer experienced an abrasive wear phenomenon [54]. This can be further confirmed by the distribution of mild and larger wear debris particles on the worn-out surface. Here, the wear tracks are not clearly visualized, which ensures the wear-resistant behaviour of the clad layer [71]. The debris removal rate was very low because the formed microstructure was rich in the γ-nickel matrix. The depth of the wear tracks was very less. This might be due to the segregation of secondary precipitates at the Colmonoy 6 clad layer. These blocky and floret precipitates have uplifted the anti-wear features at the clad layer. The blocky precipitates improved the hardness, whereas the floret precipitates enhanced the wear resistance, and the combined effect of these two precipitates has made the clad layer to exhibit excellent resistance to wear [72,73].

1.6.4 INCONEL 625 CLADDING ON INCONEL 625 SUBSTRATE

As mentioned in the cladding microstructure section (5.4), after cladding the Inconel 625 substrate with the same material, the dry sliding wear test has been carried out to interrogate the wear properties of the clad layer with the load of 40 N, sliding speed as 1 m/s, and sliding distance as 2500 m. The captured worn-out image at the Inconel

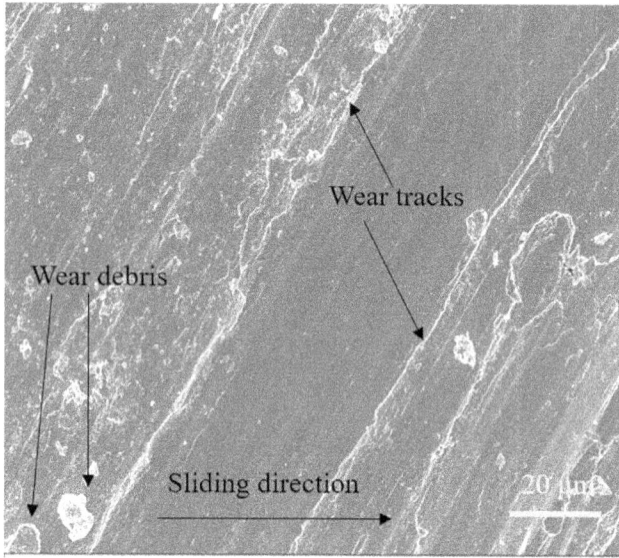

FIGURE 1.16   FESEM image of worn-out surface at the Colmonoy 6 clad layer.

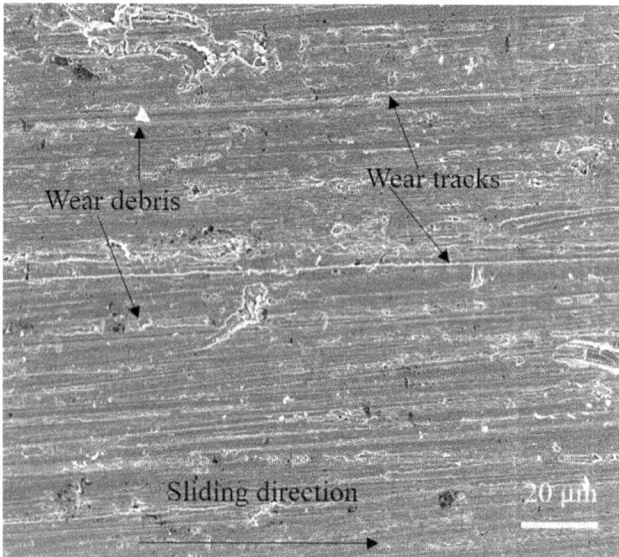

FIGURE 1.17   FESEM image of worn-out surface at the Inconel 625 clad layer.

625 clad layer is shown in Figure 1.17. From this figure, it can be observed that the wear tracks were mild with randomly distributed scratches and grooves throughout the worn-out surface. These effects were from the adhesive wear that takes place during the wear test [54]. The clad layer experienced high frictional heat that resulted

in adhesive wear. This heat creates a semi-molten wear adhesive debris that was detached from the clad surface. Considering the quantity of debris formed and wear track depth, the Inconel 625 layer has inherent resistance towards wear. This might be due to the formation of a closely packed cellular microstructure that is rich in γ-Ni dendrites along with intermetallic compounds made of nickel carbides and nickel silicide [74]. These results were observed by several literature that confirmed that Inconel 625 clad layer could resist the wear phenomena [75,76].

1.6.5 STAINLESS STEEL 420 CLADDING ON STAINLESS STEEL 410 SUBSTRATE

The dry sliding wear test was done on the stainless steel 420 clad layer with the process parameters as follows: load = 48 N; sliding speed = 0.4 m/s; sliding distance = 2500 m. Figure 1.18 shows the worn-out image taken at the clad layer. In this image, the wear debris along with the wear tracks is clearly visualized. The wear debris particles are randomly distributed along the clad layer. This distribution of wear particles showed the existence of abrasive wear phenomenon at the SS 420 clad layer [50]. The brittle phases constitute the abrasive debris but the finer wear tracks. The formation of the abrasive particles is due to the constant sliding of the tungsten ball. Somehow, the depth of the wear tracks is low, which shows that the SS 420 clad layer offers wear-resistant characteristics. This resistance is offered by the closely packed needle-like martensitic structure [77]. Several authors reported that the martensitic structure has gained wear resistance due to secondary phases that are formed after the solidification of the SS 420 clad layer [78]. These phases are rich in Fe that enhances the hardness of the clad layer, which in turn promotes maximum resistance against wear [79].

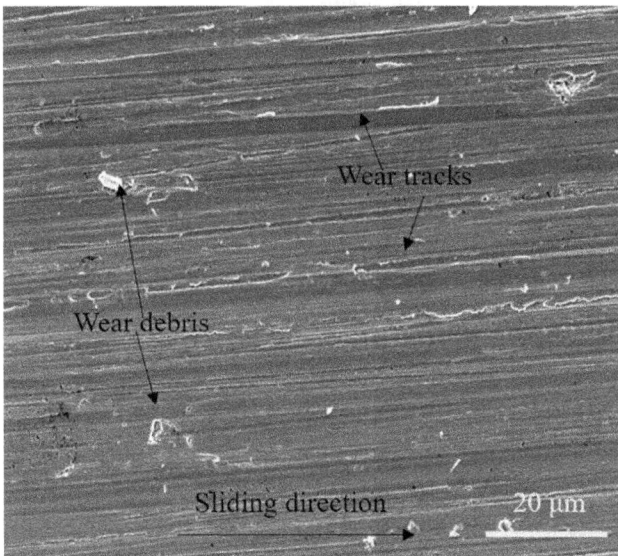

FIGURE 1.18   FESEM image of worn-out surface at the stainless steel 420 clad layer.

1.7   LASER CLADDING – HARDNESS COMPARISON

According to Archard's wear law, the hardness and wear resistance of the material are proportional to each other [80,81]. Here, the hardness property is evaluated based on the microstructure formed and correlated with the literature. The Stellite 6 coating has Cr-rich interdendritic phases. These phases consist of Cr-rich carbides that exist in different forms like Cr_7C_3, $Cr_{23}C$, and Cr_6C [46]. These carbides constitute for higher hardness at the clad layer. On the other hand, the Colmonoy claddings have secondary precipitates that got segregated after the solidification. These precipitates are rich in Cr_7C_3, $Cr_{23}C_7$, CrB, and Cr_6B_3. These borides and carbides are responsible for exhibiting excellent hardness at the Colmonoy clad layer [50,54]. The Inconel 625 and stainless steel 420 cladding offers hardness up to some extent compared to the Colmonoy coatings. Though these coatings have secondary phases, these phases are not effective when compared to the Cr-rich phases formed on the Colmonoy coatings [50]. Particularly, the Cr-carbides and Cr-borides play a significant role in enhancing the hardness and wear-resistant characteristics [54].

1.8   CONCLUDING REMARKS

This chapter discussed the wear issues in different sectors and the importance of laser cladding. The microstructure formed on the different cladded layers was investigated. The wear-resistant behaviour of these claddings was analysed with the help of worn-out SEM images, and the results were presented.

CONFLICTS OF INTEREST

The authors declare that there is no conflicts of interest.

ACKNOWLEDGEMENT

The authors like to acknowledge the editors for providing the opportunity to write this chapter.

REFERENCES

[1] Hume-Rothery W. The structure of metals and alloys. *Indian Journal of Physics.* 1969;11:74.
[2] Callister WD, Rethwisch DG. *Materials science and engineering: an introduction.* New York: Wiley; 2018 Feb 23.
[3] Hume-Rothery W. *The structures of alloys of iron: an elementary introduction.* London: Elsevier; 2016 Jan 26.
[4] Everhart J, editor. *Engineering properties of nickel and nickel alloys.* London: Springer Science & Business Media; 2012 Dec 6.
[5] Georgantzia E, Gkantou M, Kamaris GS. Aluminium alloys as structural material: a review of research. *Engineering Structures.* 2021 Jan 15;227:111372.
[6] Gepreel MA, Niinomi M. Biocompatibility of Ti-alloys for long-term implantation. *Journal of the Mechanical Behavior of Biomedical Materials.* 2013 Apr 1;20:407–15.

[7] Zmitrowicz A. Wear patterns and laws of wear–a review. *Journal of Theoretical and Applied Mechanics*. 2006;44(2):219–53.

[8] Suh NP. An overview of the delamination theory of wear. *Wear*. 1977 Aug 1;44(1):1–6.

[9] Varenberg M. Towards a unified classification of wear. *Friction*. 2013 Dec;1(4):333–40.

[10] Blau PJ. Fifty years of research on the wear of metals. *Tribology International*. 1997 May 1;30(5):321–31.

[11] Khan MA, Starr AG. Wear debris: basic features and machine health diagnostics. *Insight-Non-Destructive Testing and Condition Monitoring*. 2006 Aug 1;48(8):470–6.

[12] Swain B, Bhuyan S, Behera R, Mohapatra SS, Behera A. Wear: a serious problem in industry. In *Tribology in materials and manufacturing-wear, friction and lubrication*, edited by Patnaik A, Singh T and Kukshal V. London, UK: IntechOpen; 2020 Dec 16.

[13] Vilar R. Laser cladding. *Journal of Laser Applications*. 1999 Apr;11(2):64–79.

[14] Santo L. Laser cladding of metals: a review. *International Journal of Surface Science and Engineering*. 2008 Jan 1;2(5):327–36.

[15] Liu Y, Ding Y, Yang L, Sun R, Zhang T, Yang X. Research and progress of laser cladding on engineering alloys: a review. *Journal of Manufacturing Processes*. 2021 Jun 1;66:341–63.

[16] Reed RC. *The superalloys: fundamentals and applications*. Cambridge: Cambridge University Press; 2008 Jul 31.

[17] Geddes B, Leon H, Huang X. *Superalloys: alloying and performance*. Ohio: ASM International; 2010.

[18] Akca E, Gürsel A. A review on superalloys and IN718 nickel-based INCONEL superalloy. *Periodicals of Engineering and Natural Sciences (PEN)*. 2015 Jun 26;3(1):15–27.

[19] Weng F, Chen C, Yu H. Research status of laser cladding on titanium and its alloys: a review. *Materials & Design*. 2014 Jun 1;58:412–25.

[20] Liu J, Yu H, Chen C, Weng F, Dai J. Research and development status of laser cladding on magnesium alloys: a review. *Optics and Lasers in Engineering*. 2017 Jun 1;93:195–210.

[21] Das AK. Recent trends in laser cladding and alloying on magnesium alloys: a review. *Materials Today: Proceedings*. 2021 Jun 24;51:723–7.

[22] Singh S, Goyal DK, Kumar P, Bansal A. Laser cladding technique for erosive wear applications: a review. *Materials Research Express*. 2020 Jan 20;7(1):012007.

[23] More SR, Bhatt DV, Menghani JV. Resent research status on laser cladding as erosion resistance technique-an overview. *Materials Today: Proceedings*. 2017 Jan 1;4(9):9902–8.

[24] Arif ZU, Khalid MY, ur Rehman E, Ullah S, Atif M, Tariq A. A review on laser cladding of high-entropy alloys, their recent trends and potential applications. *Journal of Manufacturing Processes*. 2021 Aug 1;68:225–73.

[25] Quazi MM, Fazal MA, Haseeb AS, Yusof F, Masjuki HH, Arslan A. A review to the laser cladding of self-lubricating composite coatings. *Lasers in Manufacturing and Materials Processing*. 2016 Jun;3(2):67–99.

[26] Misra A, Finnie I. A review of the abrasive wear of metals. *Journal of Engineering Materials and Technology*. 1982;104:94–101.

[27] Podgornik B. Adhesive wear failures. *Journal of Failure Analysis and Prevention*. 2022 Feb;22(1):113–38.

[28] Stachowiak GW, Batchelor AW. *Engineering tribology*. Oxford: Butterworth-Heinemann; 2013 Sep 16.

[29] Affonso LO. *Machinery failure analysis handbook: sustain your operations and maximize uptime*. Houston, TX: Elsevier; 2013 Nov 25.

[30] Devaraju A. A critical review on different types of wear of materials. *International Journal of Mechanical Engineering and Technology*. 2015 Nov;6(11):77–83.

[31] Kaczorowski D, Vernot JP. Wear problems in nuclear industry. *Tribology International*. 2006 Oct 1;39(10):1286–93.

[32] Lemaire E, Le Calvar M. Evidence of tribocorrosion wear in pressurized water reactors. *Wear.* 2001 Jun 1;249(5–6):338–44.

[33] Attia MH. Fretting fatigue and wear damage of structural components in nuclear power stations—Fitness for service and life management perspective. *Tribology International.* 2006 Oct 1;39(10):1294–304.

[34] Sreenath AV, Venkatesh S. Experimental studies on the wear of engine components. *Wear.* 1970 Oct 1;16(4):245–54.

[35] Roubicek V, Raclavska H, Juchelkova D, Filip P. Wear and environmental aspects of composite materials for automotive braking industry. *Wear.* 2008 Jun 25;265(1–2):167–75.

[36] Wood RJ. Marine wear and tribocorrosion. *Wear.* 2017 Apr 15;376:893–910.

[37] Patel M, Kumar A, Pardhi B, Pal M. Abrasive, erosive and corrosive wear in slurry pumps–a review. *International Research Journal of Engineering and Technology.* 2020;7(3):2188–95.

[38] Tunna J, Sinclair J, Perez J. A review of wheel wear and rolling contact fatigue. *Proceedings of the Institution of Mechanical Engineers, Part F: Journal of Rail and Rapid Transit.* 2007 Mar 1;221(2):271–89.

[39] Hussein MA, Mohammed AS, Al-Aqeeli N. Wear characteristics of metallic biomaterials: a review. *Materials.* 2015 May 21;8(5):2749–68.

[40] Shankar V, Rao KB, Mannan SL. Microstructure and mechanical properties of Inconel 625 superalloy. *Journal of Nuclear Materials.* 2001 Feb 1;288(2–3):222–32.

[41] Paul CP, Jain A, Ganesh P, Negi J, Nath AK. Laser rapid manufacturing of Colmonoy-6 components. *Optics and Lasers in Engineering.* 2006 Oct 1;44(10):1096–109.

[42] Savanth T, Singh J, Gill JS. Laser power and scanning speed influence on the microstructure, hardness, and slurry erosion performance of Colmonoy-5 claddings. *Proceedings of the Institution of Mechanical Engineers, Part L: Journal of Materials: Design and Applications.* 2020 Jul;234(7): 947–61.

[43] D'Oliveira AS, da Silva PS, Vilar RM. Microstructural features of consecutive layers of Stellite 6 deposited by laser cladding. *Surface and Coatings Technology.* 2002 Apr 15;153(2–3):203–9.

[44] Telasang G, Majumdar JD, Padmanabham G, Manna I. Wear and corrosion behavior of laser surface engineered AISI H13 hot working tool steel. *Surface and Coatings Technology.* 2015 Jan 15;261:69–78.

[45] Candelaria AF, Pinedo CE. Influence of the heat treatment on the corrosion resistance of the martensitic stainless steel type AISI 420. *Journal of Materials Science Letters.* 2003 Aug;22(16):1151–3.

[46] Jeyaprakash N, Yang CH, Tseng SP. Wear Tribo-performances of laser cladding Colmonoy-6 and Stellite-6 Micron layers on stainless steel 304 using Yb: YAG disk laser. *Metals and Materials International.* 2021 Jun;27(6): 1540–53.

[47] Zhu Z, Ouyang C, Qiao Y, Zhou X. Wear characteristic of Stellite 6 alloy hardfacing layer by plasma arc surfacing processes. *Scanning.* 2017 Nov 20;2017:6097486.

[48] Singh R, Kumar D, Mishra SK, Tiwari SK. Laser cladding of Stellite 6 on stainless steel to enhance solid particle erosion and cavitation resistance. *Surface and Coatings Technology.* 2014 Jul 25;251:87–97.

[49] Gholipour A, Shamanian M, Ashrafizadeh F. Microstructure and wear behavior of stellite 6 cladding on 17-4 PH stainless steel. *Journal of Alloys and Compounds.* 2011 Apr 7;509(14):4905–9.

[50] Jeyaprakash N, Yang CH, Ramkumar KR, Sui GZ. Comparison of microstructure, mechanical and wear behaviour of laser cladded stainless steel 410 substrate using stainless steel 420 and Colmonoy 5 particles. *Journal of Iron and Steel Research International.* 2020 Dec;27(12):1446–55.

[51] Ming Q, Lim LC, Chen ZD. Laser cladding of nickel-based hardfacing alloys. *Surface and Coatings Technology.* 1998 Aug 4;106(2–3):174–82.

[52] Zhang H, Shi Y, Kutsuna M, Xu GJ. Laser cladding of Colmonoy 6 powder on AISI316L austenitic stainless steel. *Nuclear Engineering and Design.* 2010 Oct 1;240(10):2691–6.

[53] Paul CP, Gandhi BK, Bhargava P, Dwivedi DK, Kukreja LM. Cobalt-free laser cladding on AISI type 316L stainless steel for improved cavitation and slurry erosion wear behavior. *Journal of Materials Engineering and Performance.* 2014 Dec;23(12):4463–71.

[54] Jeyaprakash N, Yang CH, Ramkumar KR. Microstructure and wear resistance of laser cladded Inconel 625 and Colmonoy 6 depositions on Inconel 625 substrate. *Applied Physics A.* 2020 Jun;126(6):1–1.

[55] Das CR, Albert SK, Bhaduri AK, Sudha C, Terrance AL. Characterisation of nickel based hardfacing deposits on austenitic stainless steel. *Surface Engineering.* 2005 Aug 1;21(4):290–6.

[56] Balaguru S, Gupta M. Hardfacing studies of Ni alloys: a critical review. *Journal of Materials Research and Technology.* 2021 Jan 1;10:1210–42.

[57] Chakraborty G, Albert SK, Bhaduri AK. Effect of dilution and cooling rate on microstructure and magnetic properties of Ni base hardfacing alloy deposited on austenitic stainless steel. *Materials Science and Technology.* 2012 Apr 1;28(4):454–9.

[58] Jeyaprakash N, Yang CH, Prabu G, Clinktan R. Microstructure and tribological behaviour of inconel-625 superalloy produced by selective laser melting. *Metals and Materials International.* 2022 May 31;28:2997–3015.

[59] Feng K, Chen Y, Deng P, Li Y, Zhao H, Lu F, Li R, Huang J, Li Z. Improved high-temperature hardness and wear resistance of Inconel 625 coatings fabricated by laser cladding. *Journal of Materials Processing Technology.* 2017 May 1;243:82–91.

[60] Shayanfar P, Daneshmanesh H, Janghorban K. Parameters optimization for laser cladding of inconel 625 on ASTM A592 steel. *Journal of Materials Research and Technology.* 2020 Jul 1;9(4):8258–65.

[61] Abioye TE, Farayibi PK, Clare AT. A comparative study of Inconel 625 laser cladding by wire and powder feedstock. *Materials and Manufacturing Processes.* 2017 Oct 26;32(14):1653–9.

[62] Jeyaprakash N, Duraiselvam M, Raju R. Modelling of Cr_3C_2–25% NiCr laser alloyed cast iron in high temperature sliding wear condition using response surface methodology. *Archives of Metallurgy and Materials.* 2018;63:1303–15.

[63] Ma M, Wang Z, Zeng X. A comparison on metallurgical behaviors of 316L stainless steel by selective laser melting and laser cladding deposition. *Materials Science and Engineering: A.* 2017 Feb 8;685:265–73.

[64] Jeyaprakash N, Duraiselvam M, Aditya SV. Numerical modeling of WC-12% Co laser alloyed cast iron in high temperature sliding wear condition using response surface methodology. *Surface Review and Letters.* 2018 Oct 19;25(07):1950009.

[65] Birol Y. High temperature sliding wear behaviour of Inconel 617 and Stellite 6 alloys. *Wear.* 2010 Sep 17;269(9–10):664–71.

[66] Frenk A, Wagnière JD. Laser cladding with cobalt-based hardfacing alloys. *Le Journal de Physique IV.* 1991 Dec 1;1(C7):C7–65.

[67] Navas C, Conde A, Cadenas M, De Damborenea J. Tribological properties of laser clad Stellite 6 coatings on steel substrates. *Surface Engineering.* 2006 Feb 1;22(1):26–34.

[68] Hsieh CC, Chen JH, Huang FT, Wu W. Sliding wear performance of Fe-, Ni-and Co-based hardfacing alloys for PTA cladding. *International Journal of Materials Research.* 2013 Mar 14;104(3):293–300.

[69] Reinaldo PR, D'Oliveira AS. NiCrSiB coatings deposited by plasma transferred arc on different steel substrates. *Journal of Materials Engineering and Performance.* 2013 Feb;22(2):590–7.

[70] Navas C, Colaço R, De Damborenea J, Vilar R. Abrasive wear behaviour of laser clad and flame sprayed-melted NiCrBSi coatings. *Surface and Coatings Technology.* 2006 Aug 1;200(24):6854–62.

[71] Jeyaprakash N, Yang CH. Microstructure and wear behaviour of SS420 micron layers on Ti–6Al–4V substrate using laser cladding process. *Transactions of the Indian Institute of Metals.* 2020 Jun;73(6):1527–33.

[72] Deng HL, Li GL, Song YJ, Xiao SR. Microstructure and abrasion resistance mechanism of CrB particles reinforced MMC coating. In *Key engineering materials*, edited by M.K. Lei, X.P. Zhu, K.W. Xu and B.S. Xu (Vol. 373, pp. 35–38) Bäch: Trans Tech Publications Ltd; 2008.

[73] Lee K, Nam DH, Lee S. Correlation of microstructure with hardness and wear resistance in (CrB, MoB)/steel surface composites fabricated by high-energy electron beam irradiation. *Metallurgical and Materials Transactions A.* 2006 Mar;37(3):663–73.

[74] Cai LX, Wang CM, Wang HM. Laser cladding for wear-resistant Cr-alloyed Ni2Si–NiSi intermetallic composite coatings. *Materials Letters.* 2003 Jun 1;57(19):2914–8.

[75] Wang HM, Wang CM, Cai LX. Wear and corrosion resistance of laser clad Ni2Si/NiSi composite coatings. *Surface and Coatings Technology.* 2003 May 22;168(2–3):202–8.

[76] Duan G, Wang HM. High-temperature wear resistance of a laser-clad γ/Cr3Si metal silicide composite coating. *Scripta Materialia.* 2002 Jan 4;46(1):107–11.

[77] Jeyaprakash N, Yang CH, Duraiselvam M, Prabu G, Tseng SP, Kumar DR. Investigation of high temperature wear performance on laser processed nodular iron using optimization technique. *Results in Physics.* 2019 Dec 1;15:102585.

[78] Dudziński W, Konat Ł, Pękalski G. Structural and strength characteristics of wear-resistant martensitic steels. *Archives of Foundry Engineering.* 2008;8(2):21–6.

[79] Ulewicz R, Szataniak P, Novy F. Fatigue properties of wear resistant martensitic steel. In *23rd International Conference on Metallurgy and Materials* (pp. 784–789); 2014.

[80] Guo C, Zhou J, Chen J, Zhao J, Yu Y, Zhou H. High temperature wear resistance of laser cladding NiCrBSi and NiCrBSi/WC-Ni composite coatings. *Wear.* 2011 Mar 10;270(7–8):492–8.

[81] González R, Garcia MA, Penuelas I, Cadenas M, del Rocío Fernández M, Battez AH, Felgueroso D. Microstructural study of NiCrBSi coatings obtained by different processes. *Wear.* 2007 Sep 10;263(1–6):619–24.

2 Application of Corrosion-Resistant Laser Claddings

Sundara Subramanian Karuppasamy
National Taipei University of Technology

Natarajan Jeyaprakash
China University of Mining and Technology

Che-Hua Yang
National Taipei University of Technology

CONTENTS

DOI: 10.1201/9781003347408-2

2.1   INTRODUCTION

In recent years, superalloy has received huge attention in all sectors due to its notable characteristics. As the name indicates, these alloys are meant for their high performance at extreme working conditions. This alloy has high melting point and has been mostly implemented in high-temperature applications [1–3]. Mostly, these alloys are categorized into three categories: Ni-based, Fe-based, and Co-based superalloys [4]. Nickel-based superalloys are one of the most prominent superalloys that have a wide range of applications. This alloy has Ni as its major composition followed by refractory elements. It contributes to nearly 65% of the total components used in the aircraft and space shuttles. This alloy can withstand high temperature (1380°C) and deliver excellent characteristics even near to the melting point [5]. It has better tensile properties, greater strength, and good stability, and it can be predominantly used to fabricate the components used in nuclear reactors, boilers, off-shore equipment, and so on. Some of the majorly used Ni-based alloys are Inconel, Nimonic, and Colmonoy series [6]. On the other hand, cobalt-based superalloys are preferred in applications where corrosion resistance is the key aspect. This alloy has higher chromium percentage among the three categories, and cobalt can be alloyed with other elements like Fe, Si, C, etc., which had paved the way to exhibit an excellent anti-corrosion behaviour [7]. Besides the corrosion-resistant feature, this alloy is well known for its strength, toughness, processability, chemical inertness, and fabricability. These features made this alloy to get implemented in hostile environments like steam and gas turbines, pressurized water reactors, heat exchangers, propeller blades, thermal, and bio-based power plants, etc. [8]. Compared to the nickel- and cobalt-based superalloys, the iron-based superalloy has less priority and has been used in a limited number of applications. Thus, these three categories of superalloys monopolize all the sectors for many decades.

Since the superalloys are involved in fabricating the parts/components that work in harsh environments, prolonged exposure of these alloys in such environments will diminish their properties. Elevated temperature will result in the degradation of this alloy by forming hot corrosion cracks and debris [9]. In chemical industry, these alloys react with the chemicals forming salt deposits (corrosion products) containing chloride, sulphides, and sulphates. These deposits initiate the nucleation of corrosion, which render their behaviours. The oxidizing working environment will induce corrosion attacks in these alloys [10]. Thus, oxidizing environments, salt deposits, and elevated pressure and temperature are considered as the factors that influence the outstanding characteristics of these alloys [11]. In worst cases, these factors will induce cataclysmic failure of the components followed by disasters. Hence, it is required to ensure the safety of the components for the well-being of the mankind.

Corrosion is considered as the serious issue in the case of components that are made using these superalloys. It degrades the surface that initiates the formation of leak holes, voids, and other surface defects [12–14]. To protect these surfaces from corrosive media, it is better to treat the degraded surfaces via surface treatments. These treatments play an important role in protecting these surfaces from corrosion [15,16]. The surface treatments can be classified into two major categories: traditional and advanced surface treatments. Techniques like physical vapour deposition (PVD), chemical vapour deposition (CVD), coatings made via hardfacing, and thermal spray are considered as traditional techniques, whereas the laser surface treatments (LSTs) are considered as the advanced surface treatments.

In PVD process, better coating layer can be achieved by using the inorganic and organic material. But the delamination effect is unavoidable, and this technique is not cost-effective [17]. On the other hand, the CVD has higher level of environmental hazards due to the use of toxic chemicals as precursor gases [18]. Both the PVD and CVD processes can produce thin films up to 7.5 µm on the substrates. Using hardfacing technique, the damaged surfaces can be rebuilt by using hardfacing alloy powders, which might improve the resistance and lifespan of the parts. The distortion effect is considered as the major drawback of this process [19,20]. Thermal spraying involves the generation of effective coating on the substrate material. Though a wide range of substrates can get benefited by thermal spraying it with various spray materials, obtaining thick coating is very tedious. Moreover, for complex shapes (tiny and curvy surfaces), the spray efficiency is reported to be lower than the other traditional surface treatment techniques [21]. Also, most of the traditional surface treatments that are used to repair the damaged surface are quite expensive.

The limitations faced by the traditional surface treatments could be overcame by the LSTs. This technique can be implemented to materials irrespective of their geometry. It can create highly precise coatings on the substrate in a fast manner, and it is a cost-effective process [22,23]. Hence, the LSTs are mostly preferred for repairing the degraded surfaces. Laser cladding belongs to the family of LST where the substrate's surface is irradiated with the high-power laser beam. This irradiation will result in the formation of melt pool on the surface, and the cladding material is deposited on the melt pool [24]. After subsequent solidification of the melt pool, the cladded layer is obtained, which will facilitate the mechanical characteristics. Major advantages of the laser cladding process are as follows: (i) able to create fine coatings, (ii) better deposition rate, (iii) free from pores or fewer pores, (iv) low dilution and distortion effect, and (v) minimal heat-affected zone. These advantages have paved the way for the laser cladding process to treat the damaged surfaces of components [25].

For producing corrosion-resistant claddings, the choice of coating material plays a vital role in deciding the characteristics exhibited by the clad layer. Mostly, the nickel- and cobalt-based superalloys are chosen as the cladding material due to its anti-corrosion behaviour at aggressive environments [26]. These alloys have higher

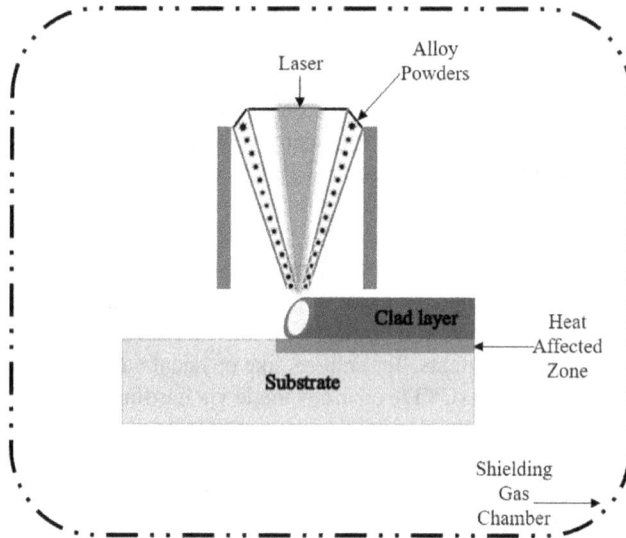

FIGURE 2.1   Schematic illustration of the laser cladding process.

chromium content, which is responsible for the formation of stable oxide film on the clad layer when exposed to corrosive media [27]. This oxide film acts as a potential barrier that avoids the intrusion of corrosion. In addition, the secondary precipitates that are formed after solidification aid in improving the mechanical behaviours like hardness, tensile strength, fatigue resistance, and other tribological properties [28]. Figure 2.1 represents the schematic illustration of the laser cladding process.

This chapter is focussed on improving the corrosion resistance of different substrates by coating them with various coating materials via laser cladding. The microstructure formed at the clad layer is investigated with the Field Emission Scanning Electron Microscope (FESEM) micrographs. The corrosion test was performed on the clad layer as per the American Society for Testing and Materials (ASTM G44-99) standards in an electrochemical workstation. The corroded morphology of the oxide film formed on the substrate and cladded layer is examined with the aid of FESEM images captured after the corrosion test. Thus, the results are presented and compared.

2.2   TYPES OF CORROSION

This section deals with the brief introduction to different types of corrosion.

2.2.1   UNIFORM CORROSION

Uniform corrosion is said to be the most common type of corrosion. When the metal surface is exposed to the corrosive media, the surface gets corroded uniformly by means of the electrochemical reaction that takes place between the corrosive medium and the metal surface. The corroded product covers relatively a larger area on the metal surface [29]. Some of the common examples are tarnishing of silver and iron rust.

2.2.2 GALVANIC CORROSION

Galvanic corrosion is defined as the corrosion process that takes place between two dissimilar metals. These metals have different electrode potentials, and in the presence of electrolyte, one metal corrodes rapidly than the other metal when they are in electrical contact [30]. This type of corrosion occurs when the metals or alloys are coated with irrelevant electrode potentials.

2.2.3 INTERGRANULAR CORROSION

Intergranular corrosion takes place at the grain boundaries of the metals. During heat treatment, the metal impurities got segregated along the grain boundaries. These impurities act as the nucleation zone for corrosion. The impurities present at the grain boundaries arrest passivation, whereas the bulk metal is free from corrosion [31]. This type of corrosion is common in heat-treated metal/alloys (e.g. ferritic chromium steels, weld decay, and austenitic Cr-Ni steels).

2.2.4 FRETTING CORROSION

Fretting corrosion happens at the contacting surface of two metals in the presence of corrosive medium under cyclic loads or vibration. During cyclic loads, the contact surface will experience some minute gaps and the corrosive medium reacts with this gap forming oxide debris [32]. This type of corrosion is noticed in clamped surfaces and in implant materials like knee and dental implants.

2.2.5 STRESS CORROSION CRACKING

This type of corrosion occurs in elevated temperature environments. The combined effect of mechanical stress and corrosive media will result in this type of corrosion. The stress may be external or internal stress. External stress can be tensile loads, and in high-temperature condition, the component experiences expansion/contraction (internal stress). Thus, these stresses will induce numerous small cracks on the surface, which act as the nucleation site for corrosion while contacting with the corrosive medium [33]. This corrosion results in sudden failure because it is hard to predict this type of corrosion. Mostly, pipelines used to transport chemical and other fluids experience stress corrosion.

2.2.6 LOCALIZED CORROSION

This type of corrosion is further divided into three categories, which are as follows:

2.2.6.1 Pitting Corrosion

Pitting corrosion is defined as the formation of small holes (pits) on the metal surface. These pits become anodic, whereas the other metal surface remains as cathode. The oxidation reaction takes place at the anodic sites and initiates the corrosion process by means of the localized galvanic reaction. Once a pit is formed, it can develop as hole or crack that penetrates deeper into the surface [34]. This type of corrosion is very hard to predict and causes sudden failure. Moreover, the presence of surface irregularities on the metal surface also leads to pitting corrosion.

2.2.6.2 Crevice Corrosion

Crevice corrosion takes place in the clogged (occluded) region of the metals in the presence of corrosive medium. Two major factors that initiate this type of corrosion are acidic environments and O_2 depletion in the clogged (crevice) region [35]. Crevice corrosion can be seen in gaskets, clamps, and couplings.

2.2.6.3 Filiform Corrosion

Filiform corrosion is usually noticed under the painted layer of Al, Mg, and steel structures. Impurities in paint will act as a passage for water to intrude in between painted layer and metal surface. In due course, the painted layer is detached from the metal surface [36]. The main factors that initiate this type of corrosion are alloy compositions, relative humidity, and temperature.

2.3 CORROSION ISSUES IN DIFFERENT SECTORS

Almost all the sectors are suffering from few types of corrosion attacks. It is estimated that the corrosion occupies nearly 5%–7% of the nation's GDP [37]. Some of the major issues are as follows: In oil and gas industries, pipelines are most often used to transport fluid substances for longer distances. These pipelines are made of steel that resists corrosion to a maximum extent. After prolonged usage, these pipelines experience intergranular corrosion and crevice corrosion. Intergranular corrosion exists in the welded area where two pipes meet, and crevice corrosion is seen in the clamps, bolt, and nut interface. The fluid medium acts as the corrosive medium, which reacts with the impurities (carbide) present in the weld region, thereby initiating intergranular corrosion at the weld joint [38]. The fuel tanks present in aerospace engines encounter corrosion attacks since the fuel contains corrosive elements that react with the tank surface and form residue that gets settled inside the tank. These residues will initiate the corrosion process that leads to the formation of leak holes followed by severe damage to the entire system [39].

In nuclear reactors, the boiler tubes experience stress corrosion attacks since these tubes transport hot steam and fluids, which will pave the way for internal tensile stress to act on these tubes. In due course, minute cracks were formed on these tubes, promoting the stress corrosion [40]. Moreover, the marine equipment reacts with the dissolved salt present in the seawater and forms salt deposits made up of chlorides and sulphides. These deposits are reported to initiate corrosion in these types of equipment [41]. Most of the medical implants are made by using bio-compatible materials like titanium alloys. These materials have excellent corrosion resistance.

But literatures reported that after few years, the hip, dental, and knee implants have corrosion issues due to pH level of blood and saliva [42]. Even the aluminium and steel structures coated with organic materials like paint exhibit filiform corrosion issues. Thus, corrosion exists in most of the sectors [43].

2.4 CORROSION-RESISTANT CLADDING POWDERS

For improving the corrosion resistance, there are certain alloy powders that are majorly preferred in laser cladding. This section deals with the brief introduction to these powders followed by the FESEM morphology of those powders.

2.4.1 INCONEL 625 POWDER

Inconel 625 alloy is one of the majorly preferred alloys for cladding the components that work in harsh environments. It is a wrought alloy consisting of major element as nickel. This alloy is solid solution-strengthened alloy, which is a combination of normal and refractory elements. This alloy is implemented in the aerospace industry from the 1950s. This alloy is well known for its properties such as yield and tensile strength, creep resistance, and sustainability in high temperatures. It has excellent corrosion resistance in NH_3 environments (heavy water reactor), and also, it provides resistance towards hot corrosion attacks [44]. This alloy forms a thick passive layer that acts as a barrier against corrosion. Some of the major applications of this alloy are as follows: pressurized water reactors, control rod, heat exchangers, and reactor core components. The as-received Inconel 625 powder particles are shown in Figure 2.2, and its average size is measured as 110 ± 35 µm. Table 2.1 lists the elemental composition of as-received Inconel 625 powder particles.

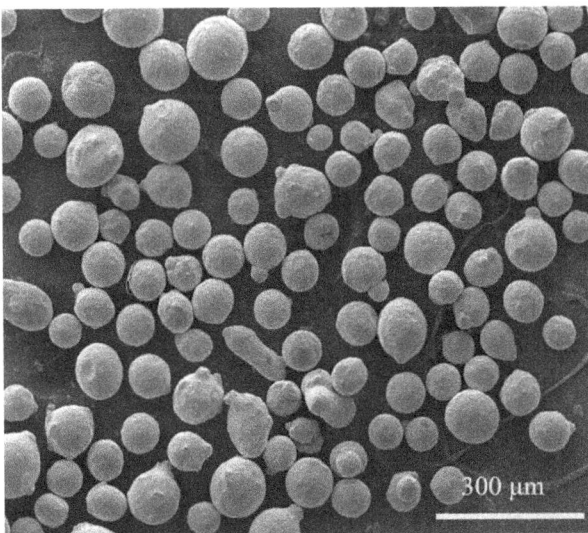

FIGURE 2.2 FESEM image of as-received Inconel 625 powder.

TABLE 2.1

Elemental Composition of Corrosion-Resistant Powder Materials

Material	Ni (%)	Co (%)	Cr (%)	Fe (%)	B (%)	Si (%)	C (%)	Mn (%)	Mo (%)	W (%)
Inconel 625	BAL	1	23	5	–	0.05	0.10	0.50	10	–
Colmonoy 6	BAL	–	14.3	4	3	4.25	0.70	–	–	–
Colmonoy 5	BAL	–	13.8	4.8	2.1	3.3	0.45	–	–	–
Stellite 6	0.62	BAL	30	0.82	–	1.27	1.22	0.14	–	4.39
H13 steel	0.3	–	5	BAL	–	1.2	0.45	0.5	1.75	–
SS 420	–	–	13.5	BAL	–	0.9	0.15	1	0.5	–

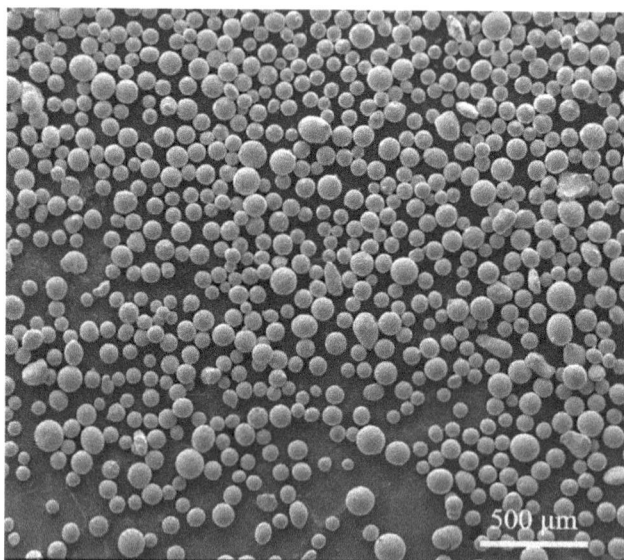

FIGURE 2.3   FESEM image of as-received Colmonoy 6 particles.

2.4.2 COLMONOY 6 POWDER

Colmonoy 6 is a type of hardfacing nickel-based superalloy. It contains significant percentages of nickel, chromium, silicon, and boron as its major constituents. This hardfacing alloy forms secondary precipitates that are rich in Ni-Cr matrix. These precipitates have uplifted the properties of Colmonoy 6 to get implemented in hostile applications. The chromium-rich boride precipitates have excellent wear resistance and hardness properties. Hence, this alloy is usually preferred for laser cladding. This alloy is used in fabricating steam and gas turbines, control rods, boilers, and components involved in thermal power plants [45]. Figure 2.3 represents the morphology of the as-received Colmonoy 6 particles with an average size of 112 ± 37 μm. The chemical composition of the as-received Colmonoy 6 particles is tabulated in Table 2.1.

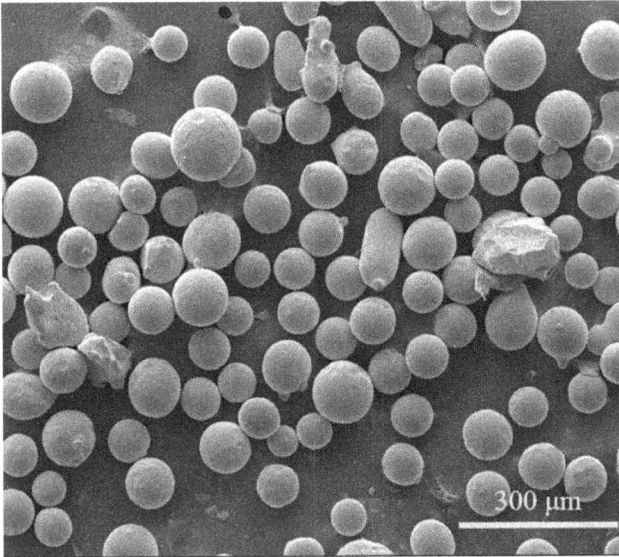

FIGURE 2.4   FESEM image of as-received Colmonoy 5 powder.

2.4.3   COLMONOY 5 POWDER

Colmonoy 5 alloy belongs to the Colmonoy series of nickel-based superalloys. It dominates the nuclear industry because of its anti-radioactive characteristics. Since the cobalt-based superalloy forms radioactive isotope (Co^{60}), there is a need for replacement of Co-based superalloy in nuclear environments. Colmonoy 5 alloy has relevant characteristics exhibited by Co-based superalloy; it can be implemented in manufacturing the components of nuclear reactors [46]. Colmonoy 5 differs from Colmonoy 6 in its chemical composition but both alloys will exhibit similar characteristics. Figure 2.4 shows the FESEM image of Colmonoy 5 used for laser cladding with an average size of 105 ± 25 µm, whereas its elemental composition is listed in Table 2.1.

2.4.4   STELLITE 6 POWDER

Stellite 6 is one of the cobalt-based superalloys, which are developed in the early 1900s. It is a combination of cobalt with chromium, tungsten, molybdenum, and other elements. This combination made this alloy to exhibit heat-resistant characteristics. This alloy has high temperature strength, hot hardness, and wear-resistant behaviours [47]. It has the ability to develop a strong oxide film made up of chromium oxide since this alloy has higher amount of chromium than other superalloys. The combined effect of oxide film and cobalt eutectics paved the way for its stability in elevated temperatures. This alloy is more suitable for cladding the damaged components like propeller shafts, turbine blades, compressors, etc. Figure 2.5 represents the FESEM image of Stellite 6 powder particles in the as-received condition. The average particle size is measured as 132 ± 40 µm, and Table 2.1 shows its chemical composition.

FIGURE 2.5   FESEM image of as-received Stellite 6 powder.

2.4.5   H13 STEEL POWDER

Laser cladding process not only uses the superalloys for improving the mechanical behaviours of the substrate. There are some metal alloys that provide better performance when laser cladded. One such alloy is the H13 steel. This steel is a hot-worked steel, which is made by combining the elements such as carbon, chromium, molybdenum, silicon, and vanadium. This steel possesses remarkable characteristics like better hardness, greater impact toughness, and excellent fatigue strength because of its strengthening elements such Mo and V (1.06%) [48]. It can withstand high thermal loads and resist corrosion. Some of the applications of this steel are die casting, plastic moulds, injection moulding, and forging components. Figure 2.6 represents the FESEM image of as-received H13 powder particles with an average size of 94 ± 26 μm, and their elemental percentage is displayed in Table 2.1.

2.4.6   STAINLESS STEEL 420 POWDER

Stainless steel 420 is a martensitic alloy, which is well known for its hardening effect, high strength, and toughness. This alloy will exhibit better performance in heat-treated conditions. It has around 13% of chromium in it. When exposed to corrosive media, this alloy forms a stable oxide film that is rich in its native oxides. After solidification, it contains a homogeneous dispersion of Cr-rich carbide phases. This phase will provide excellent hardness and anti-wear characteristics. This steel is more suitable for heat treatments since it will render improved properties at each heat-treated condition. It is used for making medical equipment, plastic moulds, and in applications where hot hardness is a key aspect [49]. Figure 2.7 shows the FESEM image of stainless steel 420 powder particles in the as-received condition. These particles have an average size of 85 ± 25 μm, and their elemental composition is listed in Table 2.1.

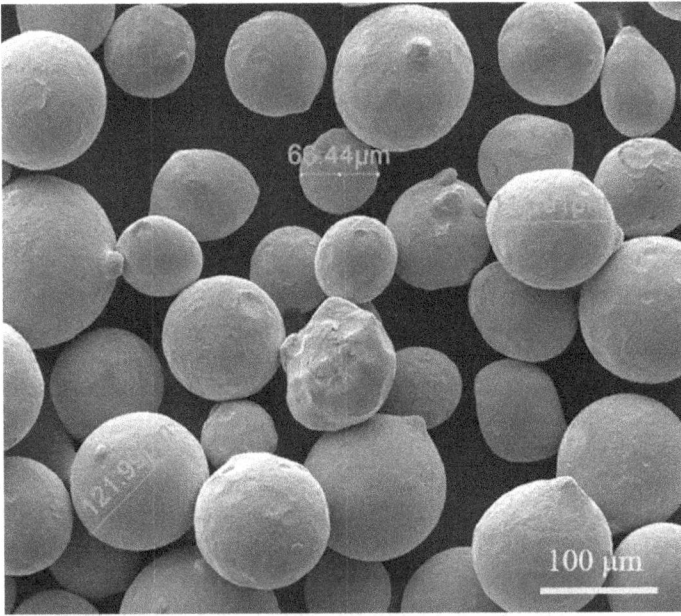

FIGURE 2.6   FESEM image of as-received H13 steel powder.

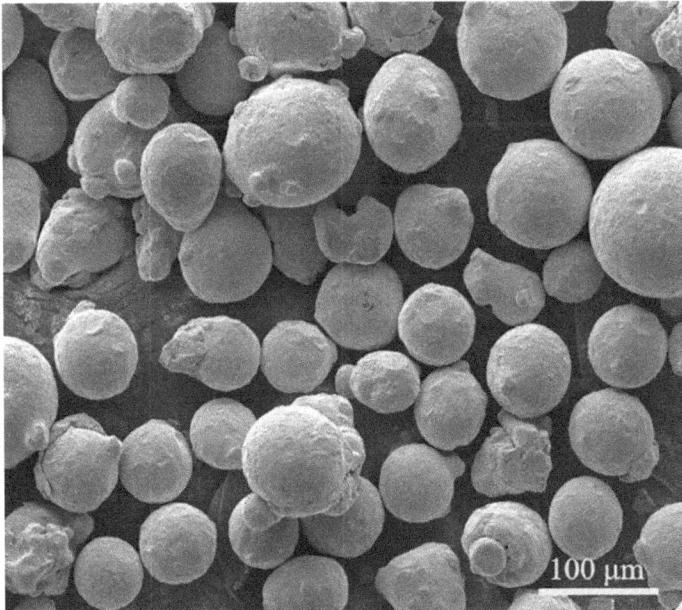

FIGURE 2.7   FESEM image of as-received stainless steel 420 powder.

2.5   LASER CLADDING – MICROSTRUCTURE EVOLUTION

This section describes the microstructure evolution that takes place during the laser cladding of different alloy powders. With the aid of FESEM images, the microstructure of the cladded layer was examined.

2.5.1   STELLITE 6 CLADDING ON INCONEL 625 SUBSTRATE

Mostly, the Inconel 625 alloy is implemented in high-temperature environments; this alloy is prone to hot corrosion attacks. Pitting corrosion was noticed in the boiler tubes of biomass-based reactors due to prolonged exposure of this alloy in hot environments. Inconel 625 readily reacts with the Cl⁻ ions that will lead to formation of chloride precipitates, which will diminish their corrosion-resistant property [50]. To improve the anti-corrosion characteristics of Inconel 625 alloy, Stellite 6 alloy is coated on it by means of laser cladding process. The process parameters that are involved in the coating process are listed in Table 2.2. After cladding, the Stellite 6 cladded samples were chemically etched with the etching solution containing 60 mL of hydrochloric acid (HCl), 15 mL of nitric, acetic acid, and distilled water. The chemically etched clad samples underwent microstructure examination using the FESEM microscope. Figure 2.8 shows the FESEM image taken at the Stellite 6 clad layer. From this figure, it can be inferred that the clad layer is rich in α-Co matrix [51]. This matrix contains both dendritic and interdendritic phases. The α-Co constitutes the dendritic eutectic phases, whereas the interdendritic areas contain the chromium-rich carbide precipitates [52]. These Cr-rich carbides are segregated as the grain boundaries of α-Co dendrites. Moreover, these carbides are reported to accelerate the tribological behaviours of the Stellite 6 clad layer [53]. Thus, the Stellite 6 clad layer has α-Co dendritic phases with Cr-carbides as their grain boundaries.

TABLE 2.2
Process Parameters Used for Laser Cladding Process

Material	Power	Feed Rate	Scanning Speed	Preheat Temperature	Shielding Gas Flow	Carrier Gas Flow
Stellite 6 on IN 625	1400 W	9 g/min	600 mm/min	150°C	25 L/min	6 SD @ 100,000/Pa
Colmonoy 5 on SS 410	1000 W	4 g/min	400 mm/min	150°C	25 L/min	6 SD @ 100,000/Pa
Colmonoy 6 on IN 625	1300 W	6 g/min	400 mm/min	100°C	25 L/min	6 SD @ 100,000/Pa
Inconel 625 on IN 625	1000 W	4 g/min	400 mm/min	150°C	25 L/min	6 SD @ 100,000/Pa
Stainless steel 420 on SS 410	1000 W	4 g/min	400 mm/min	150°C	25 L/min	6 SD @ 100,000/Pa

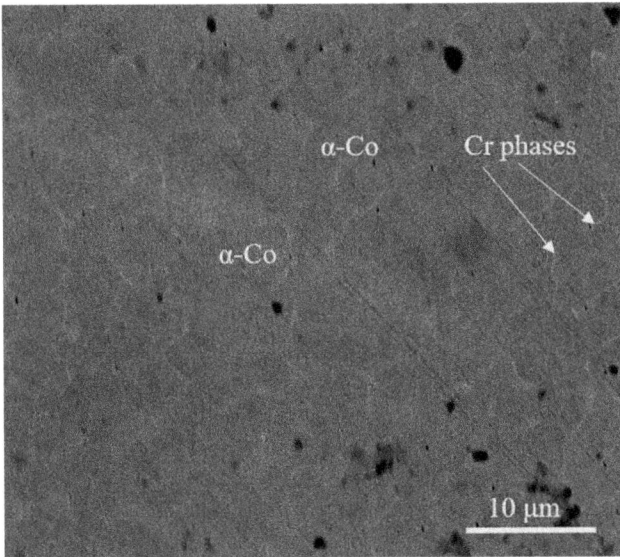

FIGURE 2.8 FESEM image of Stellite 6 clad layer.

2.5.2 COLMONOY 5 CLADDING ON STAINLESS STEEL 410 SUBSTRATE

Stainless steel 410 is used in the fabrication of hydraulic turbines, oil well tubes, etc. Pitting initiation has been reported in the hydraulic turbine blades due to the influence of Cl⁻ ions with undissolved Cr-phases present in the SS410 microstructure. These phases will act as the nucleation sites for corrosion attacks in the SS410 material [54]. Hence, the SS 410 substrate is coated with the Colmonoy 5 alloy powder, and the microstructure of the Colmonoy 5 clad layer was analysed. The laser cladding process parameters to coat the Colmonoy 5 alloy are listed in Table 2.2, and the clad samples were etched using 2% wt. H_2CrO_4 solution. From the FESEM image (Figure 2.9), the microstructure of the Colmonoy 5 clad is rich in hard laves phases. Mostly, these phases contain $(FeNi)_3B$ and Ni_3B elements [55]. Along with this primary laves phase, secondary precipitates were segregated after the solidification of the Colmonoy 5 clad layer. These precipitates were represented as dark and floret-shaped regions [56]. The dark region is made of chromium carbide, whereas the chromium borides are segregated in the floret-shaped region. The presence of hard laves along with dark and floret-shaped precipitates have uplifted the mechanical characteristics offered by the Colmonoy 5 clad layer [57].

2.5.3 COLMONOY 6 CLADDING ON INCONEL 625 SUBSTRATE

If the working environment contains Cl⁻ (marine sectors), SO_4^{2-} (fuel tanks), and F⁻ (chemical sectors), the Inconel 625 alloy will react with these ions to form dichromate compound at high temperatures. These compounds will act as the interfacial sites for pitting corrosion, which will degrade the anti-corrosion characteristics of the

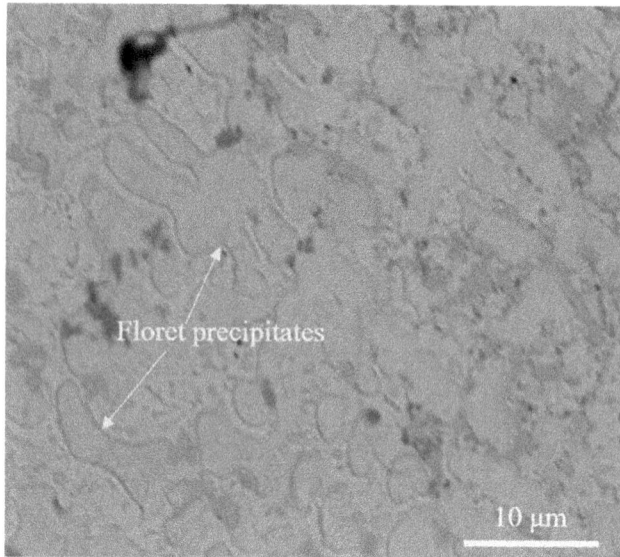

FIGURE 2.9   FESEM image of Colmonoy 5 clad layer.

Inconel 625 alloy [58]. Hence to overcome this, Inconel 625 alloy is coated with the Colmonoy 6 powder with the laser parameters listed in Table 2.2. The FESEM image of the Colmonoy 6-coated layer after etching is shown in Figure 2.10. From this figure, it can be observed that the Colmonoy 6 clad layer is rich in γ-nickel matrix since both the substrate and cladding material belong to the nickel-based superalloy family [59]. On further investigation, the γ-nickel matrix contains blocky and floret-shaped precipitates. The blocky precipitates consist of Cr_3C_2 and Cr_7C_3, whereas the floret-shaped regions are rich in chromium borides, which can be of Cr_5B_3, CrB, and Cr_2B. The CrB is reported to be formed simultaneously while interaction with the molten pool of Inconel 625, whereas the other forms are formed after the solidification [60]. The combined effect of γ-nickel matrix along with the secondary precipitates is responsible for improving the hardness and anti-wear characteristics of the Colmonoy 6 clad layer.

2.5.4   INCONEL 625 CLADDING ON INCONEL 625 SUBSTRATE

There are some applications that particularly use Inconel 625 alloy because of its higher melting point (1350°C). The corrosion attacks in such applications can be prevented by coating it with the same (Inconel 625) material. Here, the Inconel 625 substrate is coated with the same alloy via laser cladding. The optimized parameters used to coat Inconel 625 are listed in Table 2.2. The Inconel 625 cladded samples were etched with the etching solution containing 40 mL of HCl, 2 g of $CuCl_2$, and 40 mL of C_2H_5OH. After etching, the clad cross-section was analysed by means of optical micrograph (Figure 2.11). From this image, it can be seen that the Inconel

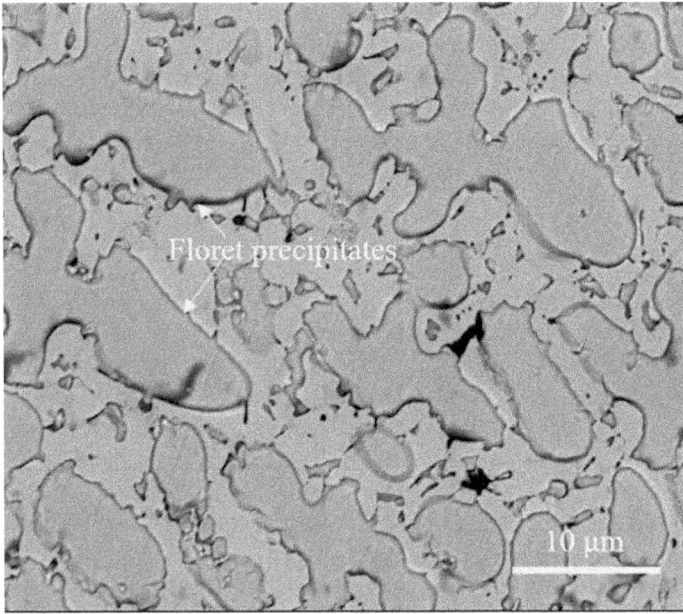

FIGURE 2.10 FESEM image of Colmonoy 6 clad layer.

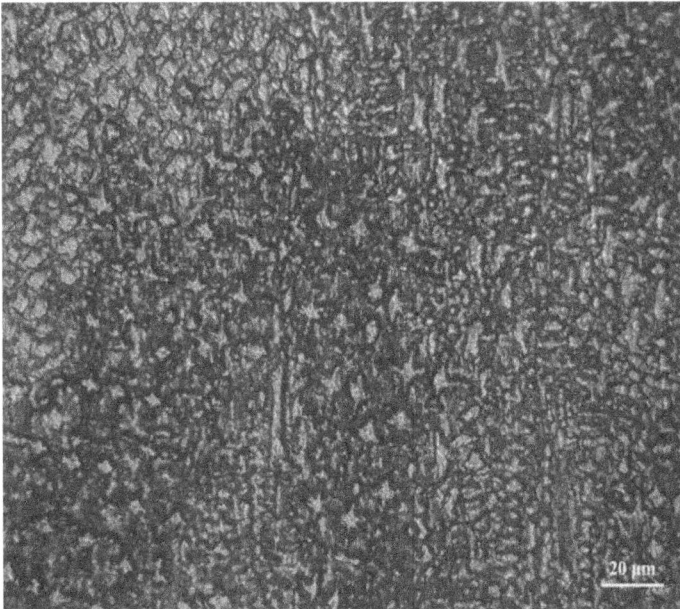

FIGURE 2.11 Optical micrograph of Inconel 625 clad layer.

625 region contains phases that are rich in γ-nickel matrix. These phases are said to be the cellular and columnar dendrites [61]. Moreover, the interdendritic precipitates got precipitated along the cellular and columnar grain boundaries. These precipitates were reported to be made of niobium carbides. After solidification, γ-nickel matrix along with laves and NbC was formed [62]. Thus, the formation of NbC along with laves helps in altering the mechanical characteristics at the clad layer [63].

2.5.5 STAINLESS STEEL 420 CLADDING ON STAINLESS STEEL 410 SUBSTRATE

The corrosion characteristics of the stainless steel 410 are heavily influenced by high temperature, molten salt deposits, and chloride ions. To enhance the corrosion resistance of the stainless steel 410, laser cladding was performed on it by using SS 420 as the cladding material. The process parameters are listed in Table 2.2, and the SS 420 clad layer is chemically etched with the solution containing 2% wt. H_2CrO_4 solution. Figure 2.12 shows the FESEM image captured after the etching process of stainless steel 420 clad layer. The microstructure of the SS 420 clad layer contains closely packed acicular needle-like morphology [64]. This morphology contributes to the martensitic phase of SS 420. Along with this phase, the intermetallic compounds were segregated along the grain boundaries of the needle phase. Researchers reported that these intermetallic compounds are rich in ferric carbides [65]. Moreover, these phases withstand high loads, and provide excellent resistance to creep, fatigue, and wear [66].

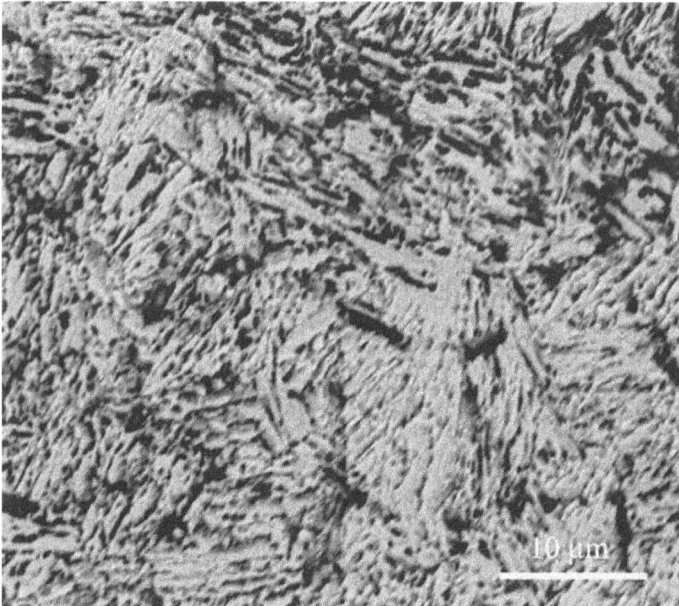

FIGURE 2.12   FESEM image of stainless steel 420 clad layer.

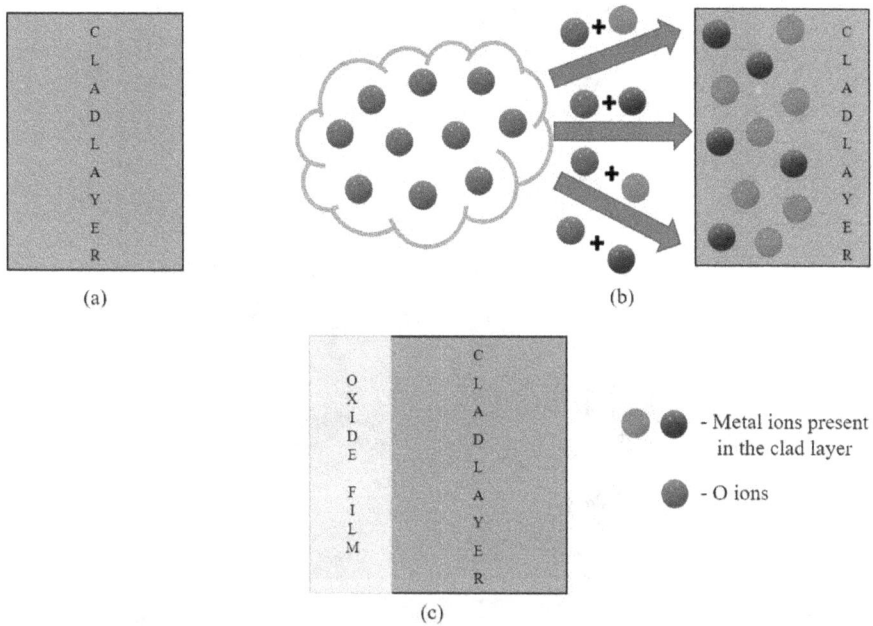

FIGURE 2.13 Schematic illustration of passive film formation on the clad layer: (a) clad layer, (b) reaction between the metal ions present in the clad layer with the oxygen ions, and (c) formation of oxide (passive film) on the clad layer which resists corrosion.

2.6 CORROSION TEST

After the cladding process, the cladded samples undergo the corrosion test (localized corrosion) for examining their corrosion-resistant characteristics. The corrosion test was carried out as per the ASTM G44–99 standards in an electrochemical setup [67]. This setup consists of three electrodes with the electrolyte. The mixture of 3.5 wt% of sodium chloride crystals with 96.5 mL of deionized water constitutes the electrolytic solution. The cladded surface serves as the testing surface, Pt wire as counter electrode, and Ag/AgCl as reference electrode [68]. This section deals with the comparison of corroded morphology of the various cladded layers along with the substrates. Figure 2.13 shows the schematic illustration of passivation film formation.

2.6.1 Stellite 6 Cladding on Inconel 625 Substrate

The corrosion test was carried out on both the Inconel 625 substrate and Stellite 6 clad samples for comparing their anti-corrosion behaviours. The corroded morphology of both samples was investigated using the FESEM images. Figure 2.14a shows the FESEM image taken at the Inconel 625 substrate after the corrosion test. It can be seen that the substrate surface has many irregularities, which denote the randomness in the oxide film formation. The oxide film formed on the substrate surface is unstable and contains pits. These pits will act as the interfacial sites that

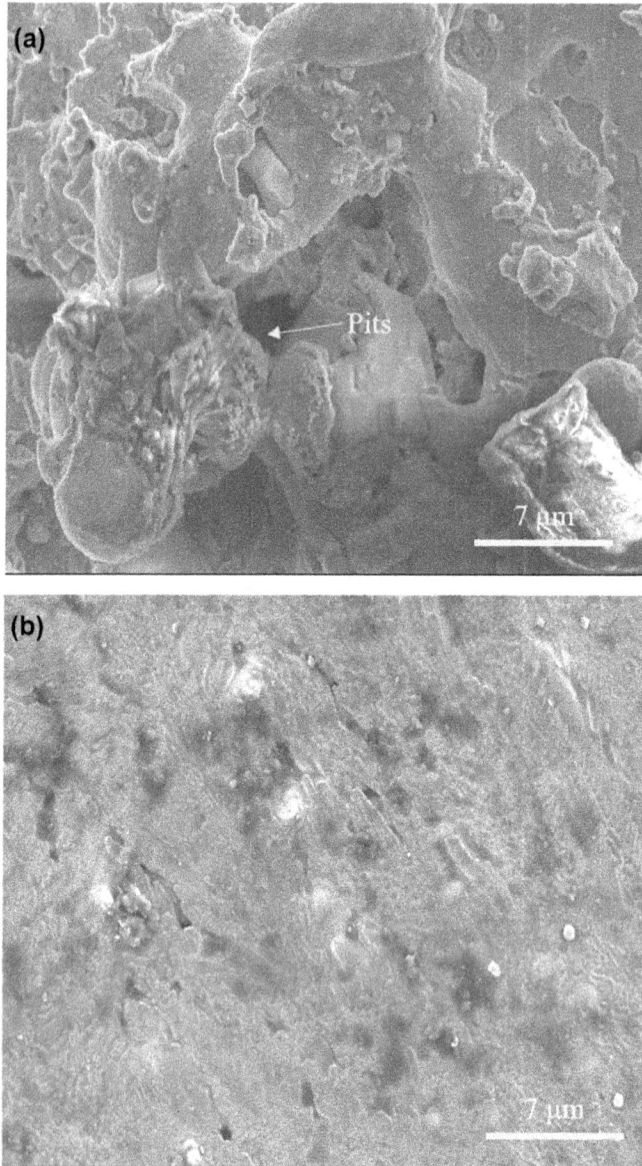

FIGURE 2.14   Passive film formed on (a) Inconel 625 substrate and (b) Stellite 6 clad layer.

initiate the corrosion. Moreover, these pits might have occurred due to the secondary precipitates that got settled as grain boundaries. Hence, these pits will minimize the corrosion resistance offered by the Inconel 625 substrate. On analysing the corroded morphology of Stellite 6 clad layer (Figure 2.14b), the oxide film tends to be continuous due to the Co-rich and Cr-rich oxides that are formed during corrosion.

Particularly, the Stellite 6 clad layer passive film is made of cobalt chromite, which is represented in the below equation.

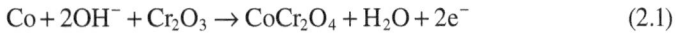

$$Co + 2OH^- + Cr_2O_3 \rightarrow CoCr_2O_4 + H_2O + 2e^- \tag{2.1}$$

This cobalt chromite ($CoCr_2O_4$) will act as the potential barrier that resists corrosion on the Stellite 6 clad layer [69,70]. It forms a stable and strong oxide film on the clad surface, thereby resisting corrosion to a maximum extent.

2.6.2 COLMONOY 5 CLADDING ON STAINLESS STEEL 410 SUBSTRATE

Stainless steel 410 resists corrosion to some extent. When exposed to corrosive media, this steel forms an oxide film on it. This film is rich in its native oxides. But in high-temperature environments, this film becomes fragile and diminishes its corrosion-resistant property. Here, the corrosion test was performed on both SS 410 substrate and Colmonoy 5 clad layer. Its corroded morphology is presented in Figure 2.15a and b. From the corroded morphology of substrate (Figure 2.15a), it can be inferred that the passive film consists of many patches of oxide layer. In other words, the formed oxide layer is supposed to be very thin, which does not withstand the corrosion potentials. Hence, it starts to break and relieve as patches during the corrosion test. In the case of Colmonoy 5 (Figure 2.15b) clad layer, the oxide film is found to be very stable [71]. This is due to the strong chromium oxide layer, which suppresses the electrochemical reaction happening at the clad surface. Equations (2.2–2.4) show the oxide film formed on the Colmonoy 5 clad layer.

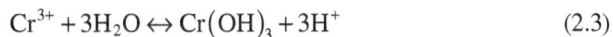

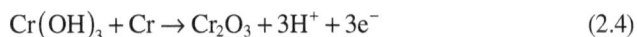

$$Cr \rightarrow Cr^{3+} + 3e^- \tag{2.2}$$

$$Cr^{3+} + 3H_2O \leftrightarrow Cr(OH)_3 + 3H^+ \tag{2.3}$$

$$Cr(OH)_3 + Cr \rightarrow Cr_2O_3 + 3H^+ + 3e^- \tag{2.4}$$

This chromium oxide was formed on the Colmonoy 5 clad layer of the stainless steel 410 substrate and resists corrosion [72].

2.6.3 COLMONOY 6 CLADDING ON INCONEL 625 SUBSTRATE

Colmonoy 6 is one of the most used hardfacing alloys, which is well known for its resistance against creep, wear, and corrosion. Here, the anti-corrosion characteristics of Colmonoy 6 cladded on the Inconel 625 substrate were examined in an electrochemical work station. In the predefined scan rate, the corrosion test was carried out on both Inconel 625 substrate and Colmonoy 6 clad layer. The oxide film formed on the both samples was analysed by means of FESEM images. Figure 2.16a represents the corroded morphology of the Inconel 625. From this figure, it can be observed that the oxide film consists of a greater number of pits. This might be due to the

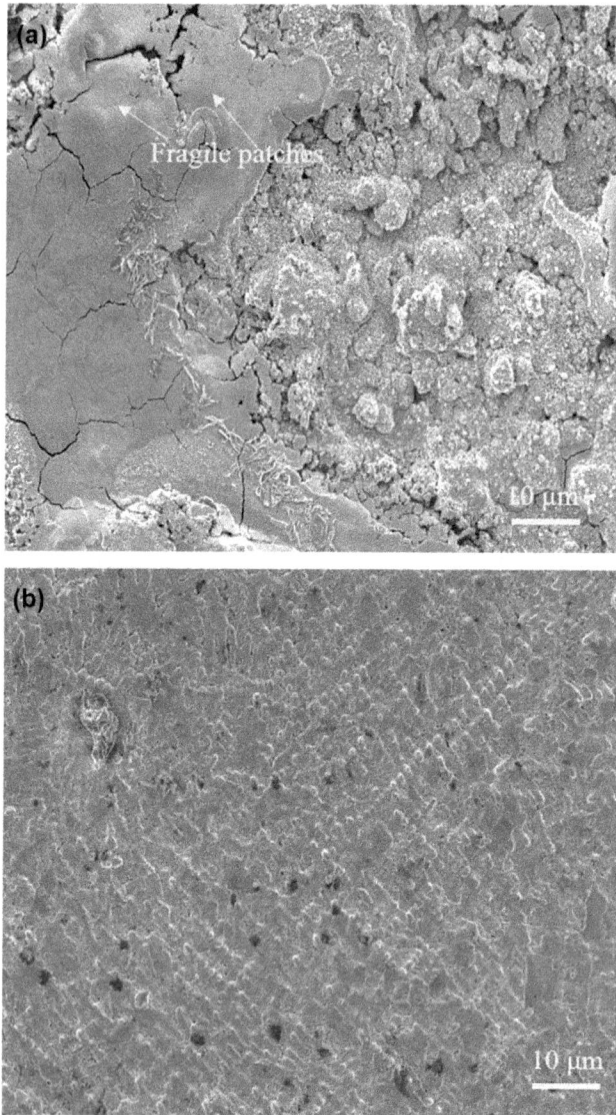

FIGURE 2.15 Passive film formed on (a) Stainless steel 410 substrate and (b) Colmonoy 5 clad layer.

impurities present at the substrate surface. These impurities act as a potential site, which initiates the pitting corrosion process. Other than the pits, the formed film tends to be uneven and unstable. The combined effect of pits and uneven film formation has resulted in the reduction of the corrosion resistance offered by the Inconel 625 alloy [73,74]. On the other hand, Colmonoy 6 clad layer exhibits better continuity in the oxide film formation (Figure 2.16b). This is because the oxide film is said to be rich in chromium oxide (Cr_2O_3), which resists corrosion to a greater extent [75].

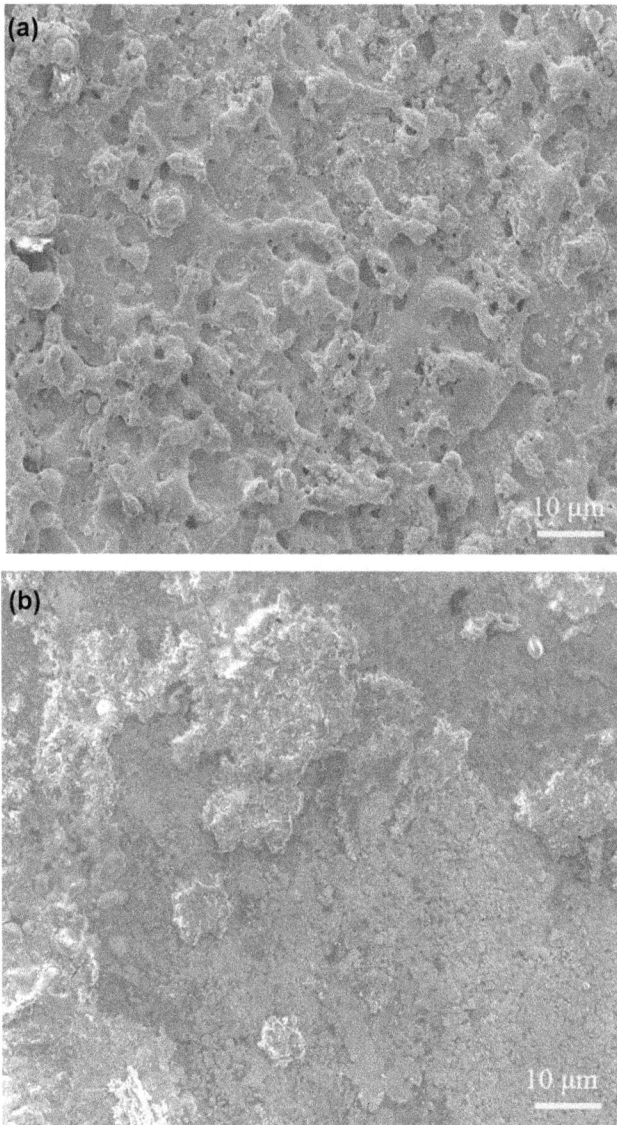

FIGURE 2.16   Passive film formed on (a) Inconel 625 substrate and (b) Colmonoy 6 clad layer.

2.6.4   INCONEL 625 CLADDING ON INCONEL 625 SUBSTRATE

Inconel 625 itself provides better anti-corrosion characteristics due to its higher percentage of chromium. Thus, the laser cladded Inconel 625 layer underwent corrosion test, and the results were compared with those for the Inconel 625 substrate. Figure 2.17a shows the FESEM corroded morphology of the Inconel 625 substrate. From this figure, it can be inferred that the passive film formed on the substrate layer

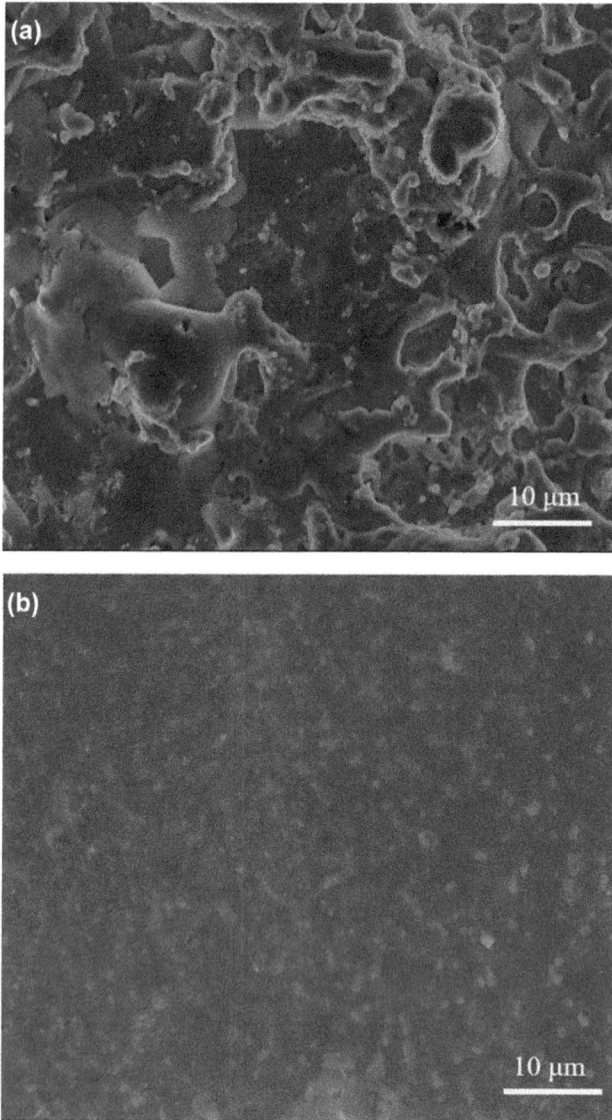

FIGURE 2.17 Passive film formed on (a) Inconel 625 substrate and (b) Inconel 625 clad layer.

tends to be fragile. More oxide patches were formed in the predefined range of scan rate and electrode potential, which shows that the substrate is prone to corrosion attacks by the Cl⁻ ions present in the electrolyte. Figure 2.17b shows the corroded morphology of the Inconel 625 clad layer. On comparing these two figures, the passive filmed formed on the clad layer is continuous, strong, and stable [76]. This is due to the effect of nickel oxide and chromium oxide that contributes to the strong passive film [77]. Equations (2.5–2.10) represent the formation of nickel oxide and chromium oxide in the clad layer [78].

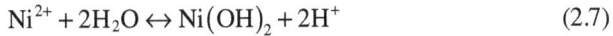

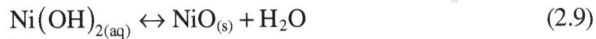

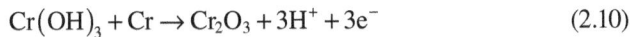

$$Ni \rightarrow Ni^{2+} + 2e^- \qquad (2.5)$$

$$Cr \rightarrow Cr^{3+} + 3e^- \qquad (2.6)$$

$$Ni^{2+} + 2H_2O \leftrightarrow Ni(OH)_2 + 2H^+ \qquad (2.7)$$

$$Cr^{3+} + 3H_2O \leftrightarrow Cr(OH)_3 + 3H^+ \qquad (2.8)$$

$$Ni(OH)_{2(aq)} \leftrightarrow NiO_{(s)} + H_2O \qquad (2.9)$$

$$Cr(OH)_3 + Cr \rightarrow Cr_2O_3 + 3H^+ + 3e^- \qquad (2.10)$$

2.6.5 STAINLESS STEEL 420 CLADDING ON STAINLESS STEEL 410 SUBSTRATE

In most cases, the stainless steel 410 components are manufactured via traditional methods like casting and forgings. The components made via these methods have undissolved Cr-precipitates, which serve as the nucleation zone for corrosion process [50]. Therefore, laser cladding has been done on the SS 410 substrate by using SS 420 as the coating material, and its corrosion resistance was compared. Figure 2.18a represents the corrosion morphology obtained on the SS 410 substrate. It can be seen that the film formed on the SS 410 substrate is fragile with a greater number of pits. The pits are due to the undissolved Cr-precipitates that got segregated on the grain boundaries [79]. This fragile oxide layer along with corrosive pits has reduced its anti-corrosion characteristics [80]. On the other hand, the passive film formed on the SS 420 clad layer (Figure 2.18b) is free from oxide patches and pits. This confirms that the SS 420 clad layer exhibits better anti-corrosion behaviour. Moreover, the clad layer is supposed to have two films. Outer thin film is composed of iron oxide, and inner thick film is composed of the chromium oxide layer [81]. The electrochemical reaction that takes place during the formation of oxide film on the SS 420 clad layer is represented in equations (2.11–2.20). Net balance equation of iron oxide is represented in equations 2.13 and 2.14, whereas equations (2.18–2.20) show the net balance reaction of chromium oxide.

$$2Fe \rightarrow 2Fe^{2+} + 2e^- \qquad (2.11)$$

$$\frac{1}{2}O_2 + H_2O + 2e^- \rightarrow 2OH^- \qquad (2.12)$$

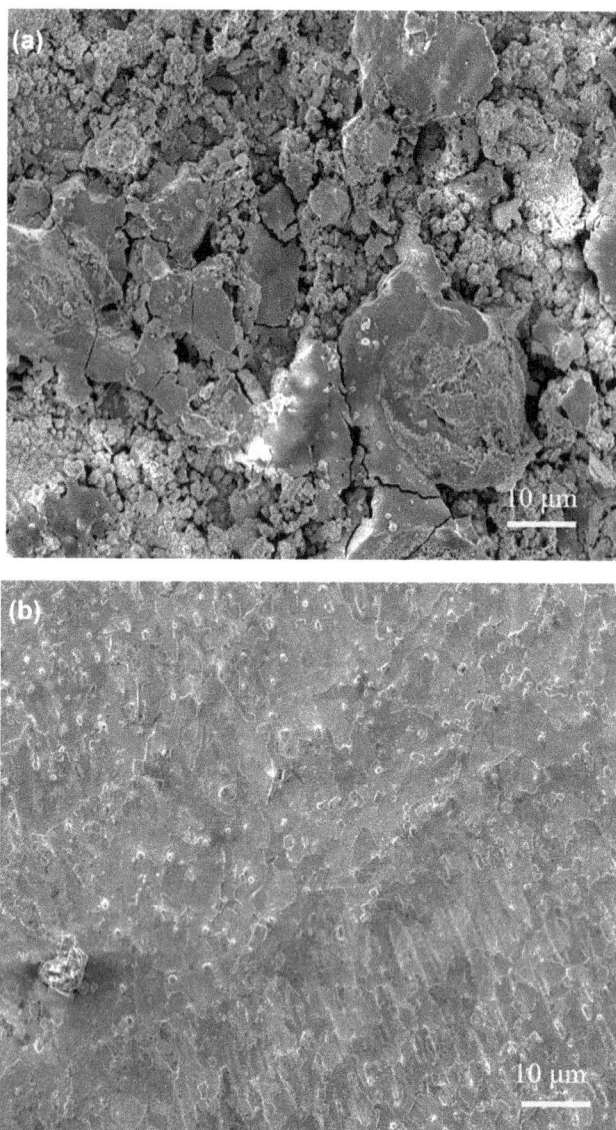

FIGURE 2.18 Passive film formed on (a) Stainless steel 410 substrate and (b) Stainless steel 420 clad layer.

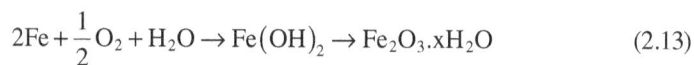

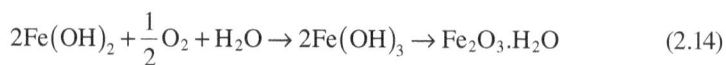

$$2Fe + \frac{1}{2}O_2 + H_2O \rightarrow Fe(OH)_2 \rightarrow Fe_2O_3.xH_2O \tag{2.13}$$

$$2Fe(OH)_2 + \frac{1}{2}O_2 + H_2O \rightarrow 2Fe(OH)_3 \rightarrow Fe_2O_3.H_2O \tag{2.14}$$

$$Cr(OH)_3 + Cr \rightarrow Cr_2O_3 + 3H^+ + 3e^- \qquad (2.15)$$

$$Cr \rightarrow Cr^{3+} + 3e^- \qquad (2.16)$$

$$Cr^{3+} + 3H_2O \leftrightarrow Cr(OH)_3 + 3H^+ \qquad (2.17)$$

$$Fe_2O_3 + 2Cr \rightarrow Cr_2O_3 + 2Fe \qquad (2.18)$$

$$3Fe_3O_4 + 8Cr \rightarrow 4Cr_2O_3 + 9Fe \qquad (2.19)$$

$$3FeO + 2Cr \rightarrow Cr_2O_3 + 3Fe \qquad (2.20)$$

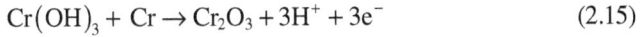

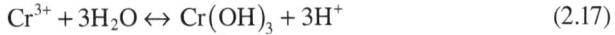

2.7 LASER CLADDING - HARDNESS COMPARISON

The superalloys are used for cladding the substrates to enhance the corrosion-resistant characteristics. Usually, the cladding process is reported to increase the hardness property [82,83]. Hence, based on the micro- and nanohardness test literatures, the following conclusion has been made. Jeyaprakash et al. [70] compared the hardness behaviour of laser cladded Stellite 6 layer with the Inconel 625 substrate. They reported that the Stellite 6 clad layer has lower indentation depth compared to the Inconel 625 substrate. This is due to the segregation of Cr-rich carbides in the Stellite 6 clad layer, which enhanced the hardness property. Karuppasamy et al. [72] analysed the hardness of Colmonoy 5 coating via microhardness and nanoindentation tests. They concluded that the Colmonoy 5 clad layer experiences lower indentation depth (nanohardness test) and higher hardness value (microhardness test) than the SS 410 substrate. The existence of Cr-rich borides and carbides in the clad layer had paved the way for improvement in the hardness property. Also, these precipitates are also reported in the Colmonoy 6 clad layer [75]. Moreover, the laser cladded Inconel 625 and stainless steel 420 showed better hardness than their respective substrates [55,59].

2.8 CONCLUDING REMARKS

This chapter presented the corrosion-resistant characteristics of various cladding powders coated on different substrates. The microstructure formed on the different clad layer has been addressed with the aid of FESEM micrographs. The anti-corrosion behaviour of the various clad layers has been investigated along with the corrosion morphologies.

CONFLICT OF INTEREST

The authors declare there is no conflicts of interest.

ACKNOWLEDGEMENT

The authors like to acknowledge the editors of this book for providing the opportunity to write this chapter.

REFERENCES

[1] Reed RC. *The superalloys: fundamentals and applications*. Cambridge: Cambridge University Press; 2008 Jul 31.

[2] Durand-Charre M. *The microstructure of superalloys*. London: Routledge; 2017 Nov 22.

[3] Geddes B, Leon H, Huang X. *Superalloys: alloying and performance*. Ohio: ASM International; 2010.

[4] Akca E, Gürsel A. A review on superalloys and IN718 nickel-based INCONEL superalloy. *Periodicals of Engineering and Natural Sciences (PEN)*. 2015 Jun 26;3(1):15–27.

[5] Jena AK, Chaturvedi MC. The role of alloying elements in the design of nickel-base superalloys. *Journal of Materials Science*. 1984 Oct;19(10):3121–39.

[6] Darolia R. Development of strong, oxidation and corrosion resistant nickel-based superalloys: critical review of challenges, progress and prospects. *International Materials Reviews*. 2019 Aug 18;64(6):355–80.

[7] Klarstrom DL. Wrought cobalt-base superalloys. *Journal of Materials Engineering and Performance*. 1993 Aug;2(4):523–30.

[8] Coutsouradis D, Davin A, Lamberigts M. Cobalt-based superalloys for applications in gas turbines. *Materials Science and Engineering*. 1987 Apr 1;88:11–9.

[9] Korashy A, Attia H, Thomson V, Oskooei S. Fretting wear behavior of cobalt-Based superalloys at high temperature–a comparative study. *Tribology International*. 2020 May 1;145:106155.

[10] Imran M, Mativenga PT, Gholinia A, Withers PJ. Comparison of tool wear mechanisms and surface integrity for dry and wet micro-drilling of nickel-base superalloys. *International Journal of Machine Tools and Manufacture*. 2014 Jan 1;76:49–60.

[11] Stringer J. High-temperature corrosion of superalloys. *Materials Science and Technology*. 1987 Jul 1;3(7):482–93.

[12] Pettit FS, Meier GH, Gell M, Kartovich CS, Bricknel RH, Kent WB, Radovich JF. Oxidation and hot corrosion of superalloys. *Superalloys*. 1984 Oct;85(1):651–87.

[13] Sidhu TS, Agrawal RD, Prakash S. Hot corrosion of some superalloys and role of high-velocity oxy-fuel spray coatings—a review. *Surface and Coatings Technology*. 2005 Aug 1;198(1–3):441–6.

[14] Hancock P. Vanadic and chloride attack of superalloys. *Materials Science and Technology*. 1987 Jul 1;3(7):536–44.

[15] Matthews A, Rickerby DS, editors. *Advanced surface coatings: a handbook of surface engineering*. Glasgow: Blackie; 1991.

[16] Davis JR, editor. *Surface engineering for corrosion and wear resistance*. Ohio: ASM International; 2001.

[17] Deng Y, Chen W, Li B, Wang C, Kuang T, Li Y. Physical vapor deposition technology for coated cutting tools: a review. *Ceramics International*. 2020 Aug 1;46(11): 18373–90.

[18] Jasinski JM, Meyerson BS, Scott BA. Mechanistic studies of chemical vapor deposition. *Annual Review of Physical Chemistry*. 1987 Oct;38(1): 109–40.

[19] Pradeep GR, Ramesh A, Prasad BD. A review paper on hardfacing processes and materials. *International Journal of Engineering Science and Technology*. 2010;2(11):6507–10.

[20] Ahn DG. Hardfacing technologies for improvement of wear characteristics of hot working tools: a review. *International Journal of Precision Engineering and Manufacturing*. 2013 Jul;14(7):1271–83.

[21] Talib RJ, Saad S, Toff MR, Hashim H. Thermal spray coating technology: a review. *Solid State Science and Technology.* 2003;11(1):109–17.

[22] Pelletier JM. Laser surface treatment: state of the art and prospects. *High-Power Lasers: Applications and Emerging Applications.* 1996 Sep 23;2789:54–64.

[23] Steen WM. Laser surface treatment: an overview. In *Laser processing: surface treatment and film deposition* (pp. 1–9); edited by Mazumder J, Conde O, Villar R, Steen W. London: Springer; 1996.

[24] Zhu L, Xue P, Lan Q, Meng G, Ren Y, Yang Z, Xu P, Liu Z. Recent research and development status of laser cladding: a review. *Optics & Laser Technology.* 2021 Jun 1;138:106915.

[25] Zhong M, Liu W. Laser surface cladding: the state of the art and challenges. *Proceedings of the Institution of Mechanical Engineers, Part C: Journal of Mechanical Engineering Science.* 2010 May 1;224(5):1041–60.

[26] Lim WY, Cao J, Suwardi A, Meng TL, Tan CK, Liu H. Recent advances in laser-cladding of metal alloys for protective coating and additive manufacturing. *Journal of Adhesion Science and Technology.* 2022 Jun 10;36(23–24):2482–504.

[27] Baldridge T, Poling G, Foroozmehr E, Kovacevic R, Metz T, Kadekar V, Gupta MC. Laser cladding of Inconel 690 on Inconel 600 superalloy for corrosion protection in nuclear applications. *Optics and Lasers in Engineering.* 2013 Feb 1;51(2):180–4.

[28] Tuominen J, Honkanen M, Hovikorpi J, Vihinen J, Vuoristo P, Maentylae T. Corrosion-resistant nickel superalloy coatings laser clad with a 6-kW high-power diode laser (HPDL). In *First International Symposium on High-Power Laser Macroprocessing* (Vol. 4831, pp. 59–64). SPIE; 2003 Mar 3.

[29] Tait WS. Controlling corrosion of chemical processing equipment. In *Handbook of environmental degradation of materials* (pp. 583–600); edited by Kutz M. Oxford: William Andrew Publishing; 2018 Jan 1.

[30] Tavakkolizadeh M, Saadatmanesh H. Galvanic corrosion of carbon and steel in aggressive environments. *Journal of Composites for Construction.* 2001 Aug 12;5(3):200–10.

[31] Tedmon CS, Vermilyea DA, Rosolowski JH. Intergranular corrosion of austenitic stainless steel. *Journal of the Electrochemical Society.* 1971 Feb 1;118(2):192.

[32] Vargel C. *Corrosion of aluminium.* Oxford: Elsevier; 2020 May 12.

[33] Raja VS, Shoji T, editors. *Stress corrosion cracking: theory and practice.* London: Elsevier; 2011 Sep 22.

[34] Akpanyung KV, Loto RT. Pitting corrosion evaluation: a review. In *Journal of Physics: Conference Series* (Vol. 1378, No. 2, p. 022088). IOP Publishing; 2019 Dec 1.

[35] Rashidi N, Alavi-Soltani SR, Asmatulu R. Crevice corrosion theory, mechanisms and prevention methods. In *Proceedings of the 3rd Annual GRASP Symposium* (pp. 215–216). Wichita State University; 2007.

[36] Bautista A. Filiform corrosion in polymer-coated metals. *Progress in Organic Coatings.* 1996 May 1;28(1):49–58.

[37] Koch G. Cost of corrosion. Trends in oil and gas corrosion research and technologies. *Elsevier.* 2017 Jan 1;2017:3–30.

[38] Cheng YF. Pipeline corrosion. *Corrosion Engineering, Science and Technology.* 2015 May 1;50(3):161–2.

[39] Benavides S. Corrosion in the aerospace industry. In *Corrosion control in the aerospace industry* (pp. 1–14); edited by Samuel Benavides. Boston, MA: Woodhead Publishing; 2009 Jan 1.

[40] Cattant F, Crusset D, Féron D. Corrosion issues in nuclear industry today. *Materials Today.* 2008 Oct 1;11(10):32–7.

[41] Wang Z, Zhou Z, Xu W, Yang D, Xu Y, Yang L, Ren J, Li Y, Huang Y. Research status and development trends in the field of marine environment corrosion: a new perspective. *Environmental Science and Pollution Research.* 2021 Oct;28(39): 54403–28.

[42] Manivasagam G, Dhinasekaran D, Rajamanickam A. Biomedical implants: corrosion and its prevention-a review. *Recent Patents on Corrosion Science*. 2010 May 24;2(1), 40–54.

[43] Steele GD. Filiform corrosion on architectural aluminium: a review. *Anti-Corrosion Methods and Materials*. 1994;41(1), 8–12.

[44] Shankar V, Rao KB, Mannan SL. Microstructure and mechanical properties of Inconel 625 superalloy. *Journal of Nuclear Materials*. 2001 Feb 1;288(2–3):222–32.

[45] Ming Q, Lim LC, Chen ZD. Laser cladding of nickel-based hardfacing alloys. *Surface and Coatings Technology*. 1998 Aug 4;106(2–3):174–82.

[46] Bhaduri AK, Indira R, Albert SK, Rao BP, Jain SC, Asokkumar S. Selection of hardfacing material for components of the Indian Prototype Fast Breeder Reactor. *Journal of Nuclear Materials*. 2004 Sep 1;334(2–3):109–14.

[47] Sassatelli P, Bolelli G, Gualtieri ML, Heinonen E, Honkanen M, Lusvarghi L, Manfredini T, Rigon R, Vippola M. Properties of HVOF-sprayed Stellite-6 coatings. *Surface and Coatings Technology*. 2018 Mar 25;338:45–62.

[48] Wang J, Liu S, Fang Y, He Z. A short review on selective laser melting of H13 steel. *The International Journal of Advanced Manufacturing Technology*. 2020 Jun;108(7): 2453–66.

[49] Candelaria AF, Pinedo CE. Influence of the heat treatment on the corrosion resistance of the martensitic stainless steel type AISI 420. *Journal of Materials Science Letters*. 2003 Aug;22(16):1151–3.

[50] Kim H, Mitton D, Latanision RM. Stress corrosion cracking of alloly 625 in Ph 2 aqueous solution at high temperature and pressure. In *Corrosion 2010*. Texas: OnePetro; 2010 Mar 14.

[51] Jeyaprakash N, Yang CH, Tseng SP. Wear Tribo-performances of laser cladding Colmonoy-6 and Stellite-6 Micron layers on stainless steel 304 using Yb: YAG disk laser. *Metals and Materials International*. 2021 Jun;27(6):1540–53.

[52] Zhu Z, Ouyang C, Qiao Y, Zhou X. Wear characteristic of Stellite 6 alloy hardfacing layer by plasma arc surfacing processes. *Scanning*. 2017 Nov 20;2017:6097486.

[53] Houdková Š, Pala Z, Smazalová E, Vostřák M, Česánek Z. Microstructure and sliding wear properties of HVOF sprayed, laser remelted and laser clad Stellite 6 coatings. *Surface and Coatings Technology*. 2017 May 25;318:129–41.

[54] Taji I, Moayed MH, Mirjalili M. Correlation between sensitisation and pitting corrosion of AISI 403 martensitic stainless steel. *Corrosion Science*. 2015 Mar 1;92:301–8.

[55] Jeyaprakash N, Yang CH, Ramkumar KR, Sui GZ. Comparison of microstructure, mechanical and wear behaviour of laser cladded stainless steel 410 substrate using stainless steel 420 and Colmonoy 5 particles. *Journal of Iron and Steel Research International*. 2020 Dec;27(12):1446–55.

[56] Gnanasekaran S, Padmanaban G, Balasubramanian V, Kumar H, Albert SK. Laser Hardfacing of Colmonoy-5 (Ni-Cr-Si-BC) powder onto 316LN austenitic stainless steel: effect of powder feed rate on microstructure, mechanical properties and tribological behavior. *Lasers in Engineering (Old City Publishing)*. 2019 Jul 1;42:283–302.

[57] Jeyaprakash N, Yang CH, Tseng SP. Characterization and tribological evaluation of NiCrMoNb and NiCrBSiC laser cladding on near-α titanium alloy. *The International Journal of Advanced Manufacturing Technology*. 2020 Jan;106(5):2347–61.

[58] Wang L, Li H, Liu Q, Xu L, Lin S, Zheng K. Effect of sodium chloride on the electrochemical corrosion of Inconel 625 at high temperature and pressure. *Journal of Alloys and Compounds*. 2017 May 5;703:523–9.

[59] Jeyaprakash N, Yang CH, Ramkumar KR. Microstructure and wear resistance of laser cladded Inconel 625 and Colmonoy 6 depositions on Inconel 625 substrate. *Applied Physics A*. 2020 Jun;126(6): 1–11.

[60] Hemmati I, Ocelík V, De Hosson JT. Effects of the alloy composition on phase constitution and properties of laser deposited Ni-Cr-B-Si coatings. *Physics Procedia*. 2013 Jan 1;41:302–11.

[61] Chaudhuri A, Raghupathy Y, Srinivasan D, Suwas S, Srivastava C. Microstructural evolution of cold-sprayed Inconel 625 superalloy coatings on low alloy steel substrate. *Acta Materialia*. 2017 May 1;129:11–25.

[62] Xu X, Mi G, Chen L, Xiong L, Jiang P, Shao X, Wang C. Research on microstructures and properties of Inconel 625 coatings obtained by laser cladding with wire. *Journal of Alloys and Compounds*. 2017 Aug 25;715:362–73.

[63] Huebner J, Kusiński J, Rutkowski P, Kata D. Microstructural and mechanical study of inconel 625–tungsten carbide composite coatings obtained by powder laser cladding. *Archives of Metallurgy and Materials*. 2017;62(2):531–38.

[64] Jeyaprakash N, Yang CH. Microstructure and wear behaviour of SS420 micron layers on Ti–6Al–4V substrate using laser cladding process. *Transactions of the Indian Institute of Metals*. 2020 Jun;73(6):1527–33.

[65] Saqib SM, Urbanic RJ. Investigation of the transient characteristics for laser cladding beads using 420 stainless steel powder. *Journal of Manufacturing Science and Engineering*. 2017 Aug 1;139(8):081009.

[66] Da Sun S, Fabijanic D, Barr C, Liu Q, Walker K, Matthews N, Orchowski N, Easton M, Brandt M. In-situ quench and tempering for microstructure control and enhanced mechanical properties of laser cladded AISI 420 stainless steel powder on 300M steel substrates. *Surface and Coatings Technology*. 2018 Jan 15;333:210–9.

[67] Jeyaprakash N, Kantipudi MB, Yang CH. Study on microstructure and anti-corrosion property of stainless steel particles deposition on Ti–6Al–4V substrate using laser cladding technique. *Lasers in Manufacturing and Materials Processing*. 2022 Jun;9(2):214–27.

[68] Jeyaprakash N, Yang CH, Susila P, Karuppasamy SS. Laser cladding of NiCrMoFeNbTa particles on Inconel 625 alloy: Microstructure and corrosion resistance. *Transactions of the Indian Institute of Metals*. 2022 Nov 1;1–4.

[69] Behazin M, Biesinger MC, Noël JJ, Wren JC. Comparative study of film formation on high-purity Co and Stellite-6: probing the roles of a chromium oxide layer and gamma-radiation. *Corrosion Science*. 2012 Oct 1;63:40–50.

[70] Jeyaprakash N, Yang CH, Karuppasamy SS, Dhineshkumar SR. Evaluation of microstructure, nanoindentation and corrosion behavior of laser cladded Stellite-6 alloy on Inconel-625 substrate. *Materials Today Communications*. 2022 Jun 1;31:103370.

[71] La Barbera-Sosa JG, Santana YY, Villalobos-Gutiérrez C, Cabello-Sequera S, Staia MH, Puchi-Cabrera ES. Effect of spray distance on the corrosion-fatigue behavior of a medium-carbon steel coated with a Colmonoy 88 alloy deposited by HVOF thermal spray. *Surface and Coatings Technology*. 2010 Nov 15;205(4):1137–44.

[72] Karuppasamy SS, Jeyaprakash N, Yang CH. Microstructure, nanoindentation and corrosion behavior of Colmonoy-5 deposition on SS410 substrate using laser cladding process. *Arabian Journal for Science and Engineering*. 2022 Jan 6;47:8751–67.

[73] Brooking L, Sumner J, Gray S, Simms NJ. Stress corrosion of Ni-based superalloys. *Materials at High Temperatures*. 2018 May 4;35(1–3):120–9.

[74] Osoba LO, Oladoye AM, Ogbonna VE. Corrosion evaluation of superalloys Haynes 282 and Inconel 718 in hydrochloric acid. *Journal of Alloys and Compounds*. 2019 Oct 5;804:376–84.

[75] Jeyaprakash N, Yang CH, Karuppasamy SS. Laser cladding of Colmonoy 6 particles on Inconel 625 substrate: microstructure and corrosion resistance. *Surface Review and Letters*. 2022 May 20;29(8):1–5.

[76] Scendo M, Staszewska-Samson K, Danielewski H. Corrosion behavior of Inconel 625 coating produced by laser cladding. *Coatings*. 2021 Jun 24;11(7):759.

[77] Fesharaki MN, Shoja-Razavi R, Mansouri HA, Jamali H. Evaluation of the hot corrosion behavior of Inconel 625 coatings on the Inconel 738 substrate by laser and TIG cladding techniques. *Optics & Laser Technology*. 2019 Apr 1;111:744–53.

[78] Schmuki P, Virtanen S, Davenport AJ, Vitus CM. Transpassive dissolution of Cr and sputter-deposited Cr oxides studied by in situ X-ray near-edge spectroscopy. *Journal of the Electrochemical Society*. 1996 Dec 1;143(12):3997.

[79] Agrawal R, Namboodhiri TK. The inhibition of sulphuric acid corrosion of 410 stainless steel by thioureas. *Corrosion Science*. 1990 Jan 1;30(1):37–52.

[80] Jeyaprakash N, Yang CH, Karuppasamy SS, Rajendran DK. Correlation of microstructural with corrosion behaviour of Ti-6Al-4V specimens developed through selective laser melting technique. *Proceedings of the Institution of Mechanical Engineers, Part E: Journal of Process Mechanical Engineering*. 2022 Mar 22;236(5):2240–51.

[81] Luo X, Tang R, Long C, Miao Z, Peng Q, Li C. Corrosion behavior of austenitic and ferritic steels in supercritical water. *Nuclear Engineering and Technology*. 2008;40(2):147–54.

[82] Davim JP, Oliveira C, Cardoso A. Laser cladding: an experimental study of geometric form and hardness of coating using statistical analysis. *Proceedings of the Institution of Mechanical Engineers, Part B: Journal of Engineering Manufacture*. 2006 Sep 1;220(9):1549–54.

[83] Paydas H, Mertens A, Carrus R, Lecomte-Beckers J, Tchuindjang JT. Laser cladding as repair technology for Ti–6Al–4V alloy: influence of building strategy on microstructure and hardness. *Materials & Design*. 2015 Nov 15;85:497–510.

3 A Comprehensive Analysis of Recent Advancements and Future Prospects in Laser-Based Additive Manufacturing Techniques

Amrinder Mehta, Hitesh Vasudev, and Chander Prakash
Lovely Professional University

Sharanjit Singh
DAV University

Kuldeep K. Saxena
Lovely Professional University

CONTENTS

3.1   INTRODUCTION

The term "additive manufacturing" refers to a group of relatively new techniques for making things by adding layers of material in rapid succession [1–4]. The production of completely functional metal parts can now be accomplished using additive manufacturing (AM), despite the fact that the potential applications of AM have expanded in recent years [5]. There is a variety of surface development techniques such as thermal spray coatings, cladding, nitriding, cyaniding, carburizing, etc. [6–37]. Newer surface development techniques are beneficial in terms of development of a versatile surface that can serve in erosion, corrosion, abrasion, and biodegradation processes [33–35,38–72]. Metal AM is being utilized in some of the most demanding areas of the manufacturing industry, including aerospace, energy, defence, and biomedicine (AM) [58,69–71]. It was previously impossible to make items with intricate, topologically optimal structures and inside holes using normal manufacturing methods; however, with the help of this technology, it is now viable to do so. However, using AM technologies by themselves is often insufficient to meet the stringent tolerance and surface integrity requirements that are imposed. The manufacturing of net-shaped models of geometrically complex items is made possible by AM technologies. These technologies are founded on sliced CAD models of the solid portions that are sought. In later years, research was carried out on the powder-based AM techniques of directed energy deposition (DED) and laser powder bed fusion (LPBF). Post-processing is typically necessary for additively manufactured parts in order for them to attain the desired level of surface smoothness, dimensional tolerances, and mechanical quality standards. So, it is not surprising that hybrid AM, which is the idea of combining AM and post-processing technologies into single- and multiple-setup processing solutions, has become a popular idea in the industry and is getting a lot of attention in terms of research and development. Despite the versatility and unique benefits of AM technologies, there is a requirement for a unified structure that incorporates methods for fabricating unique material with innovative features. This framework is needed even though AM technologies have these benefits [73].

As a consequence of this, the goal of this review is to use a bottom-up methodology to close the gap between what powder-based AM metallic samples are made of and how they behave. It is required to customize the precursor materials, which in this case are the elemental powders, in order to achieve ideal microstructures and properties. The first thing that has to be done in order to make powder-based AM is to get the feedstock powder ready. The AM process can use either a mixture of component particles to form the final product, which can then be supplied to the machine [74]. When compared to the production of alloyed components using pre-alloyed powders, the development of AM techniques for making alloyed components from elementally mixed mixtures is more cost-effective. It provides efficient composition tailoring, which opens up a wealth of opportunities and avenues for developing new alloy mixtures and formulated quality functionally graded. Accurate design control is among the most essential components of any effective in situ alloying procedure. This part of the procedure is mostly determined by the feedstock utilized and the preparation method that is utilized. The enthalpy of mixing, the Marangoni force, and the phenomena known as the "columnar equivalent transition" (CET) are three more key variables [75].

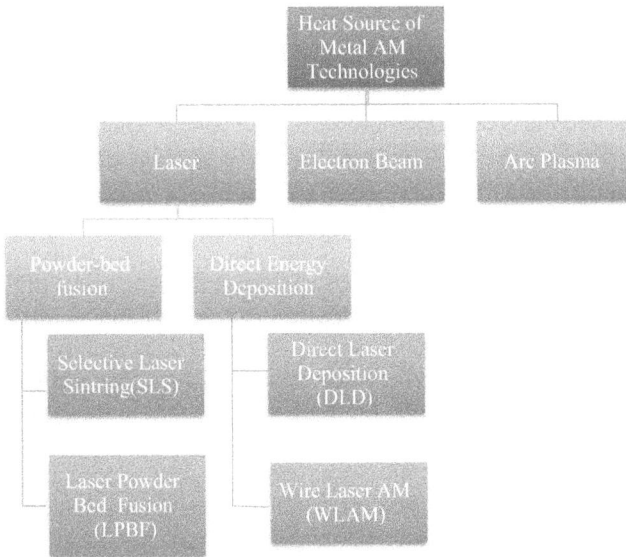

FIGURE 3.1   Classification of metal additive manufacturing.

The diagram illustrates the classification of existing AM technologies for producing metal components (Figure 3.1). Researchers can get an idea of the processes being examined in this evaluation by looking at the accompanying picture. As was said earlier, the primary emphasis of this study is placed on single-setup and multi-setup hybrid AM techniques that make use of an AM process that is based on the utilization of a laser to mix metal powders.

This study investigates the AM efforts on alloy production by utilizing elemental powders as feedstock [76]. Additionally, the processes that take place throughout in situ alloying of AM specimens are investigated by keeping the industry's perspective on target, i.e. the constraints and potential benefits of employing the design and production technique in AM.

3.2   METHODS FOR (AM)-BASED ON FUSION

Fusing the material, which is usually in the form of wire or powder, may now be done in a variety of different methods thanks to advancements in technology. A laser, an electron beam, can all function as thermal source, and these three techniques are among the most often utilized ones. DED and powder bed fusion (PBF), which is also known as "plasma bed fusion", are the two principal ways that have been used. Both of these procedures have proven successful DED [77–92]. This study focuses on AM processes that make use of powder rather than wire because powder-based AM techniques have been put into practice a great deal more frequently than wire-based AM techniques [93]. Powder bed and powder feed AM methods are referred to by their initials, LPBF and DED, respectively.

3.2.1   LASER POWDER BED FUSION (LPBF) TECHNIQUES

However, when the LPBF technology was still in its infancy, the density and strength of the manufactured part were inadequate for the application because of insufficient powder fusion and the created part's sensitivity to powder spheroidization afterward melting [94]. The incorporation of fibre lasers with superior efficiency and the optimization of the manufacturing process have resulted in a significant improvement in the making accuracy, and mechanical qualities of titanium alloys, superalloys, and aluminium alloys produced by the LPBF method. These improvements have been achieved through a combination of factors. Since that time, the LPBF technique has made steady progress toward the top of the list of commercialized AM technologies in a variety of industries, including the medical, automotive, and aerospace industries. Figure 3.2a illustrates the steps involved in the LPBF technology process. Firstly, the recoater blade covers the substrate with a thin metallic powder coat or earlier existing layers. Next, scanning for selective point-by-point irradiation is carried out with the laser beam in accordance with the shape of the pieces' two-dimensional cross-sections at a specific rate [95]. This makes it possible to melt the radioactive metal powder. As the laser beam moves away, the liquid metal and powders swiftly cool and solidify into their respective forms. After that, the height of the construction platform will be decreased by the amount that was chosen in accordance with the thickness of the coating. Repeating the technique described above up until the point where the full component is built is required [96]. To create a dense and faultless component, it is essential that the procedure characteristics, like scanning speed and laser power, be compatible with the powder substance and layer thickness of the powder. The entire LPBF process is normally carried out in an environment that is inert in order to prevent oxidation at high temperatures, with an average particle size of 40 μm. Although it does have substantial limitations, the LPBF technique is usually used to make very minor and accurate parts because of its low build efficiency and remarkable dimensional precision [97]. Despite these benefits, the technology still has considerable restrictions. During the LPBF process, it can be difficult to prevent the powder from spheroidizing, which leads to the formation of small pores. This is a challenge that must be overcome.

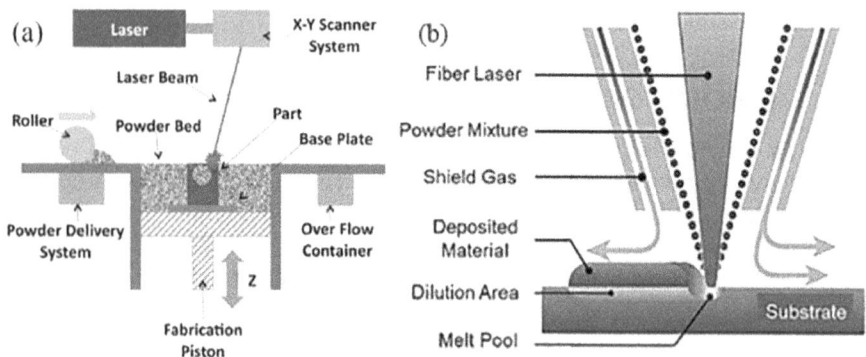

FIGURE 3.2   Diagrammatic illustration of the (a) LPBF and (b) LDED technique [95,98].

3.2.2 PROCESSES INVOLVING LASER-DIRECTED ENERGY DEPOSITION (LDED)

The technology known as "laser-directed energy deposition" (LDED) was independently developed by a large number of research organizations located all over the world. Laser solid forming, laser metal deposition, laser-designed net shaping, and so on are just some of the terms that are used for this process, despite the fact that the fundamental concepts are, for the most part, the same. The technical workings of LDED that is based on powder are shown in Figure 3.2b. LDED discretizes the three-dimensional (3D) model into layers of two dimensions (2D), just like LPBF does [98]. However, in contrast to LPBF, LDED allows the user to choose whether the feedstock materials consist of wire, powder, or both. The components of the additive are delivered into the melt pool rather than being dispersed across a powder bed. The LDED technology makes use of a higher laser power and a larger laser beam size than the LPBF technology does in order to achieve a greater build efficiency than the latter. Furthermore, LDED can be put to excellent use in the repair of high-performance and high-value components, as well as in the manufacture of gradient structures through the synchronous feeding of a variety of materials [99]. Both of these applications are examples of excellent uses for LDED. However, the usage of LDED technology is rather limited because it is difficult to make anything with very intricate shapes using the technology.

The great majority of LDED processes include directly feeding a melt pool that has been previously formed a thermal source on the material with metallic powders through nozzles (Figure 3.3) [100]. This is done in almost all cases. Although it is possible to utilize a powder that has already been alloyed as the feed material, or a combination of different elemental powders, each of which can be produced via its own nozzle if so desired [101]. Because of this, the engineering composition of the finished alloy can be modified to meet the particular requirements of the region that is currently being worked on. In contrast to LDED technologies that rely on powder as the feedstock, which unavoidably lead to the loss of some of the blown powders

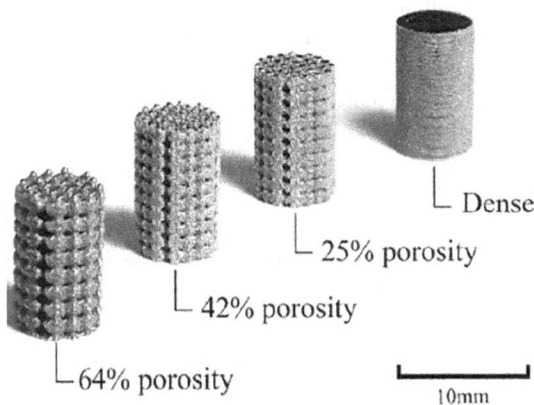

L Dense

L 25% porosity

L 42% porosity

L 64% porosity

10mm

FIGURE 3.3 Explanation graphically of hard and porous Ti-Ta cylinders manufactured by the LPBF process beginning with elemental particles [100].

due to the nature of the powders themselves, certain LDED methods make use of wire as the feedstock. In spite of the fact that these techniques are less popular, the end result is a greater deposition efficiency and rate as a result of this. Products that are wire-fed, on the other hand, have a surface that is more uneven and, as a result, call for a more thorough polishing process. The most significant benefits of utilizing LDED procedures are that they enable the production of Functionally graded materials (FGM), the repair of engineering goods that have intricate shapes, and the cladding of exteriors with particularly fabricated metal alloys [102]. These are the three types of applications that have the potential to benefit the most from the utilization of LDED. By shifting the building stage in relation to the nozzles, which are typically fixed in place, which may be accomplished with the help of a CNC table, the diversity of LDED methods can be expanded even further. This enables the creation of designs that are more intricate.

3.3   PROCESS PARAMETERS

When dealing with fusion-based AM processes, the most important process factors to take into consideration are the layer thickness, the hatching distance, the laser power (P), and the laser scan rate (V). Both the LPBF and the LDED methods make use of a number of different process parameters in their respective analyses. Both the feedstock and the laser will be provided concurrently while the LDED process is being carried out. In order to successfully melt the metallic particles, a slower laser scan rate and a more powerful laser are utilized in this process. This is because these factors enable the production of a larger melt pool, which is required for the process to be successful. Instead, the LPBF method produces a fully dense part from stationed powders by utilizing greater scan speeds, finer particles, and lower laser intensities [103]. This allows for a more efficient use of the material. This method demands a lower amount of total inputted energy compared to previous ways. In addition, it has been established that the LDED technique and the LPBF procedure both result in different solidification microstructures due to the varied melting processes and heat transmission modes that are utilized. This is the case because both of these procedures are utilized to solidify the material. With in situ alloying in mind, the LPBF process is much more delicate to the structure and distribution of particle sizes of the feedstock nanoparticle, which in turn affects the flow properties of the composition [104]. In other words, the shape and distribution of the feedstock particles' sizes have an effect on the flowability of the mixture. The ability of the mixture to flow is impacted by both of these considerations. Because maintaining a consistent composition requires that the powder layer be distributed in a manner that is uniform over the building platform, it is necessary that the powder combination be able to be followed concerning this particular point. Nevertheless, in accordance with the LDED method, the flowability of the pressure of the carrier gas has a considerable amount of control over the feedstock [105]. In the great majority of LPBF operations, pre-alloyed powders are utilized as an input material. On the other hand, the ultimate composition of some of the research projects was achieved by employing the LPBF technique and utilizing elemental powder combinations as the means of production. In spite of this fact, elemental powders are utilized in the LDED method a great deal more regularly. This is owing to the fact that the approach provides a greater level of control over the chemical composition as a result of the capacity to independently change the flow of powder from each individual nozzle. This is the primary reason for the success of the method. In addition to providing a wider melt pool and lasers with a higher power output, this approach also provides increased control over the chemical makeup of the final product. In addition to that, using this function will allow you to make gradient compositions of your own. When the LPBF process is used to do in situ alloying, it is possible for the components to experience something that is known as the "unmelted particle-keyhole". Although it is common knowledge that LPBF processes make it possible to create finer features, it is also possible for the components to experience this phenomenon [106]. The issue is that significantly more energy is needed in order to melt the particles thoroughly and the development of keyholes at the expense of the density of the material.

3.4 PROCESS OPTIMIZATION

It is vital to pick the size and quality of the constituents by the attributes of the alloying elements to obtain a final product that is dense and consistent in appearance. Because of this, it will be possible to produce a component that is dense and uniform across its entirety. According to the research [107], the three most important physical characteristics affect the powder feedstock's superiority. There is a connection between the quality of the powder feedstock and these three parameters given above [107]. The seeming density of the feedstock is substantially impacted by various factors, some of which include the characteristics that were discussed earlier, for example. There is a decreased possibility that there will be resolved unmelted particles when smaller particles are employed, and the overall energy absorption is increased as a result. Finer particles in a powder combination that has a wide size distribution curve have the potential to increase the friction between the particles, which in turn decreases the feedstock's flowability. This can happen if the powder has an extremely broad-size bimodal distribution. Additionally, better particles can fill the places that are left behind by larger particles, which results in an apparent increase in density. This is because finer particles have a greater surface area-to-volume ratio [108]. There is more information available concerning the manner in which the physicochemical qualities of the constituent powders have an influence on the features of the completed feedstock or product. In AM in situ alloying, some of the essential thermophysical characteristics that must be taken into consideration include the melting point and an individual particle's reflectiveness. In point of fact, more care needs to be taken to guarantee that the feedstock is correctly generated in accordance with the characteristics indicated earlier. A non-uniform flow pattern in elemental blends that comprise particles of varied densities and shapes may, as a side effect, yield an erroneous alloy composition [109,110]. This is because the powders have different densities and forms. When powder mixtures contain materials that have significantly different laser absorption capabilities, the macro-segregation of the constituent powders has a significant impact on the local energy absorption of the powder combination as a whole. This is because macro-segregation causes the constituent powders to become more or less homogeneous. The melting and solidifying processes of the

AM samples are influenced as a result of this factor. It is reasonable to anticipate that those whose constituents have melting points that are noticeably different will run into unmelted particles. This is due to the fact that the melting points of the components will be very different from one another. Homogeneity refers to the degree to which the constituents are mixed together in the same way throughout the sample. This is because the viscosity and density of the liquid state is a measure of how fluid the constituents are. On the other hand, if the heat source is insufficient to melt the heavier component totally, density will become a more critical consideration [111]. However, before moving further with the production of alloys that can be utilized in AM procedures, researchers frequently take into account a number of other factors first. AM processes, in contrast to more conventional methods of production, have primary essential properties such as directional thermal gradients and quick cooling-heating cycles. AM is distinguished from more conventional manufacturing processes by virtue of these qualities.

The rough textured microstructures that are produced by AM processes have their origin in the strong temperature gradient that exists within the melt pool. In Figure 3.4, the relationship between parameters, thermal behaviour, and quality is depicted. The arrows point out numerous critical aspects that relate to thermal transfer, such as temperature gradient, cooling rate, and heat input. Moreover, the arrows also represent the quality of the thermal behaviour [112]. In a wide variety of applications, the presence of these microstructures is not desirable. If the substrate is preheated, then the severity of the roughness can be lessened to some extent; nevertheless, this strategy is not feasible in the majority of situations. Another strategy that can be utilized to lessen the effects of the temperature gradient by altering the crystal structure of the material is one method for accomplishing this goal. When carrying out the AM process, some alloying elements, most notably W, Mo, and Nb, are expected to have a role in grain refining, as hypothesized by the theory that is based on phase diagrams. One is able to make an educated judgement regarding the degree to which the alloying element will work well as a nucleant by using the phase diagram, which represents the atomic properties of the elements [113]. This allows one to

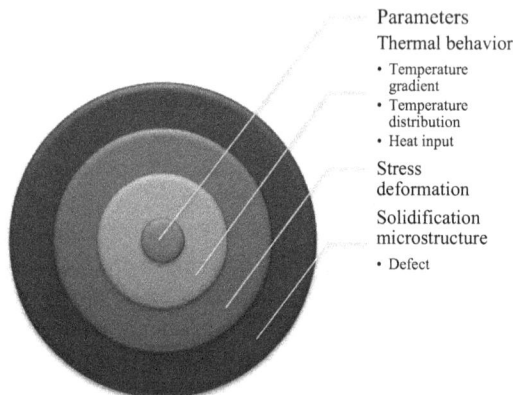

Parameters
Thermal behavior
• Temperature gradient
• Temperature distribution
• Heat input
Stress deformation
Solidification microstructure
• Defect

FIGURE 3.4   Relation between parameters, thermal behaviour, and quality.

make an accurate prediction. In earlier attempts at AM, the growth restriction factor (Q) was utilized to determine an element's effectiveness. This was done in order to determine whether or not an element was suitable for the AM process. In order to determine whether or not the element might activate the nuclei, this step was taken. This highlights how the existence of a solute element can alter the rate at which a diffusion boundary layer arises in a system that, by its very nature, is undercooled.

3.5 UNIFORMITY OF COMPOSITION IN CONVENTIONAL AND MECHANICAL MIXING

Mixing the powder is the first step that must be completed in an in situ AM process. This step is necessary because it ensures that the final alloy will have a composition that is uniform over the whole piece. In addition, solidification happens not long after laser scanning, and it is followed by melting, both during laser scanning and actions that take place while the material is in the solid state. The Marangoni effect is one example of an additional component that, at this stage, has the potential to influence the compositional uniformity [114]. During the phase of the process referred to as "solidification", the nucleation rate, grain size, solute segregation, and laser rescanning of the solidified tracks are among the most important factors. Last but not least, subsequent heat treatment to AM can improve the homogeneity of alloys that were created in situ. The next part provides a succinct overview of several aspects of the first and third processes, such as the mixing of powder and the product's subsequent solidification. It is difficult to provide a concise summary of the elements that influence the other phases because of how closely they are related to one another. Because of this, some of these elements are mentioned throughout this book. This is because it is difficult to categorize these elements into distinct themes due to how closely they are related to one another. In parts that were deposited from elemental powders that were mixed in the conventional manner, it is possible for chemical segregation of the alloy components to take place [115]. This phenomenon is known as "chemical segregation". The key contributors to this segregation in the initial feedstock are variations in the density, shape, and size of the constituent powders. These variations play a role in causing the segregation. This indicates how crucial it is to carry out the elemental powder combination assembly process in accordance with the prescribed steps. This is the case despite the fact that uniformity of the alloy can be improved by changing the settings of the laser process. It would seem that in situ alloying using AM will inevitably result in the production of chemical segregation at some point in the process. The initial heterogeneity of the feedstock is the most important aspect that plays a role in determining the overall outcome [116]. The manner in which the constituents are dispersed throughout the melt pool is affected not only by the Marangoni force, which helps to ensure that the constituents of the alloy are thoroughly mixed, but also by the melting point and viscosity of the individual elements. The Marangoni force helps to ensure that the constituents of the alloy are thoroughly mixed. Mixing elemental powders often involves using tube shakers, because this machine needs the least amount of energy and provides the most efficient mixing. This helps to keep the amount of macro-segregation that takes place to a manageable level [117]. In situations like this, it is optimal to make use of mental powders that have a narrow PSD in order to bring the level of inhomogeneity in the feedstock down to a more manageable level. It has been demonstrated through previous research that the degree to which the powder feedstock is mixed ought to be related to the size of the beam spot in order to accomplish the level of chemical and microstructural uniformity that was aimed. This was one of the goals that needed to be accomplished. If the beams were more focused, as those used in low-pressure beam forming (LPBF), then it would be even more important to make sure that the necessary amount of feedstock mixing was taking place [118].

The majority of the time, an increase in the powder's energy is produced as a direct result of the mechanical mixing of the particles. This is true regardless of the type of powder being worked with. As the foundational material for the micro-DED of NiTi alloys, we utilized elemental mixtures of non-spherical Ti and Ni powders that were processed mechanically using a milling machine. This element was essential in the production of the NiTi alloys. They arrived at the conclusion that mechanical alloying possesses the ability to stimulate the expansion of the NiTi phase. This was an important step in the process. This led to a spike in the laser scan rate of three times its previous level, which, in turn, led to an increase in the probability of the melt pool forming [119]. In contrast to the production of undesirable intermetallic phases, the formation of the desirable equiatomic NiTi phase is achieved. Because samples produced with optimized LPBF do not typically have such a high level of porosity, the scientists stated that the primary cause of the porosity was created throughout the ball milling method. This was because the porosity was created when the powder was ground up by the balls. The process of ball milling was responsible for the formation of the porosity, which explains why this is the case. Ball milling is a pre-processing operation that can be used to change the morphology of powders that are of an irregular shape and give the necessary flowability for AM techniques. Producing the necessary flowability is the means through which this can be accomplished. Ball milling will be used to produce almost spherical stainless steel powders in order to meet the powder quality requirements for AM [120]. This will be done in order to obtain the appropriate powder characteristics. This will make it possible for the powders to have the qualities we want. Using the process of ball milling, non-spherical particles of hydrogenated-dehydrogenated titanium (HDH-Ti) can be made to take on a shape that is virtually spherical.

In this way, the larger parent particles can be embellished with one or more types of tiny decorative particles. This is an innovative approach that was developed only not too long ago for the purpose of mixing the AM feedstock (Figure 3.5). According to studies, the procedure is able to increase the uniformity of the AM's feed-source while still maintaining the same cost advantages and recyclability as previous mixing methods [121]. This is in contrast to the fact that the procedure has the potential to increase the level of uniformity [122,123]. It was possible to carry out the Ti-, Al-based alloy by utilizing metallic powders that were synthesized by using either the straightforward mixing of the elemental powders or the satelliting of the elemental powders. Both of these methods were successful in producing the elemental powders (Figure 3.5). The findings of the investigation indicate that satelliting results in the production of structural properties that are equivalent to specimens that are

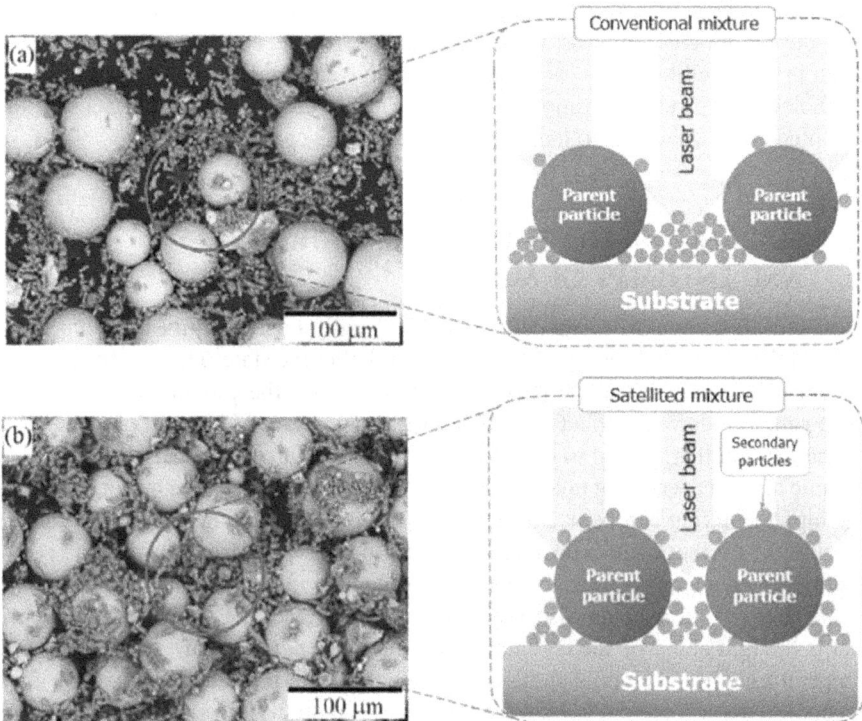

FIGURE 3.5 Backscattered SEM pictures depict the distribution of Ti, Al, and V in the (a) blended mixture and (b) satellited powders; smaller Al and V particles adorn bigger Ti particles [121].

deposited from Ti-6Al-4V. This is due to the fact that satelliting has an effect on the energy absorption and, as a direct consequence of this, the melting and solidification modes [121]. In spite of the fact that inhomogeneities were found in the samples that were produced by using a satellited feedstock, satelliting was successful in reducing the amount of chemical segregation that took place in comparison with the samples that were deposited from conventional combinations of elemental powders [124]. For the satellization process to be carried out successfully, it is crucial to pay attention to the morphology of the powders, the size variation between the basic and linked particles, and the liquid mixing of the powders. Only then can this process be carried out to its full potential. In addition to the benefits that were outlined before, the satelliting tactic also includes a few drawbacks that should be taken into consideration. For example, the sizes of the parent and connected particles must be sufficiently dissimilar to one another in order to prevent the smaller particles from acting as carriers and exclusively latching on to the larger parent particles. This can be accomplished by ensuring that the sizes of the parent and connected particles are sufficiently dissimilar to one another. In addition, it is vital to emphasize that minuscule particles that are prepared for satelliting are not always easily accessible. This is something that needs to be emphasized very strongly [125,126]. In addition, there is a possibility that, during the DED process, the pressure that is provided by the carrier gas could result in the satellited particles detaching, which would reduce the usefulness of the technology for direct deposition processes. This would be due to the fact that the satellited particles would no longer be attached to the carrier gas.

3.6 MARANGONI EFFECT IN SOLIDIFICATION AND ENTHALPY OF MIXING

In in situ alloying, AM methods are used to locally melt and mix the component elemental powders in order to obtain the desired alloying state. This is done in order to achieve the intended alloying state. Because of this, the enthalpy of mixing has the potential to have an effect not only on the temperature of the melt pool, but also on the method that is used to carry out in situ alloy synthesis. When there are exothermic mixing events that take place in conjunction with a negative mixing heat, the enthalpy of mixing can be useful in the production of a homogenous alloy. This is because the exothermic mixing reactions drive the heat away from the mixture [127]. The melting pool may come into contact with a temperature gradient as a direct result of the energy source being discussed. On the other hand, the temperature increase that is brought about by the enthalpy of mixing does not have any influence on the temperature of the melt pool at all since it only has an effect when the reaction zone is conditioned adiabatically. However, a number of studies have come to the same conclusion, which is that an exothermic reaction may help to cut down on the amount of laser energy that is used and increase the procedure's energy efficiency when it comes to either LPBF or LDED [128].

This conclusion has been reached as a result of the fact that an exothermic reaction has the potential to help. The orientation of the grains in a layer that has recently been deposited on top of another layer is typically determined by the orientation of the grains in the layer that was deposited on top of that layer. When a DLD sample has a cubic crystal structure, the grains develop in the direction of the heat flow and in the crystallographic direction, respectively. This occurs when the grain growth is driven by crystallographic orientation. This takes place whenever the sample is subjected to multiple cycles of heating and cooling [130]. On the other hand, the EBSD examination of the DLD sample revealed anisotropy despite the absence of any visibly evident texture features, as seen in Figure 3.6a, c, and e. This result may be seen in the figure. Because of the method of reciprocating scanning that was utilized in this study, the direction of the heat flux was muddled, and as a result, it was not possible to see the microstructure in any particular direction [129]. This was due to the fact that the study was unable to see the microstructure in any particular direction. As indicated by the red lines in Figure 3.6b, d, and f, bainite can contain both low-angle grain boundaries and high-angle grain boundaries inside its crystal structure (represented by black lines). In order to construct thermal modelling for the LPBF method utilizing the Ti-B combination, in which the high temperature generated by the reaction was taken into theoretical consideration as a potential source of energy, the LPBF process needed to use the Ti-B combination [129]. This was accomplished

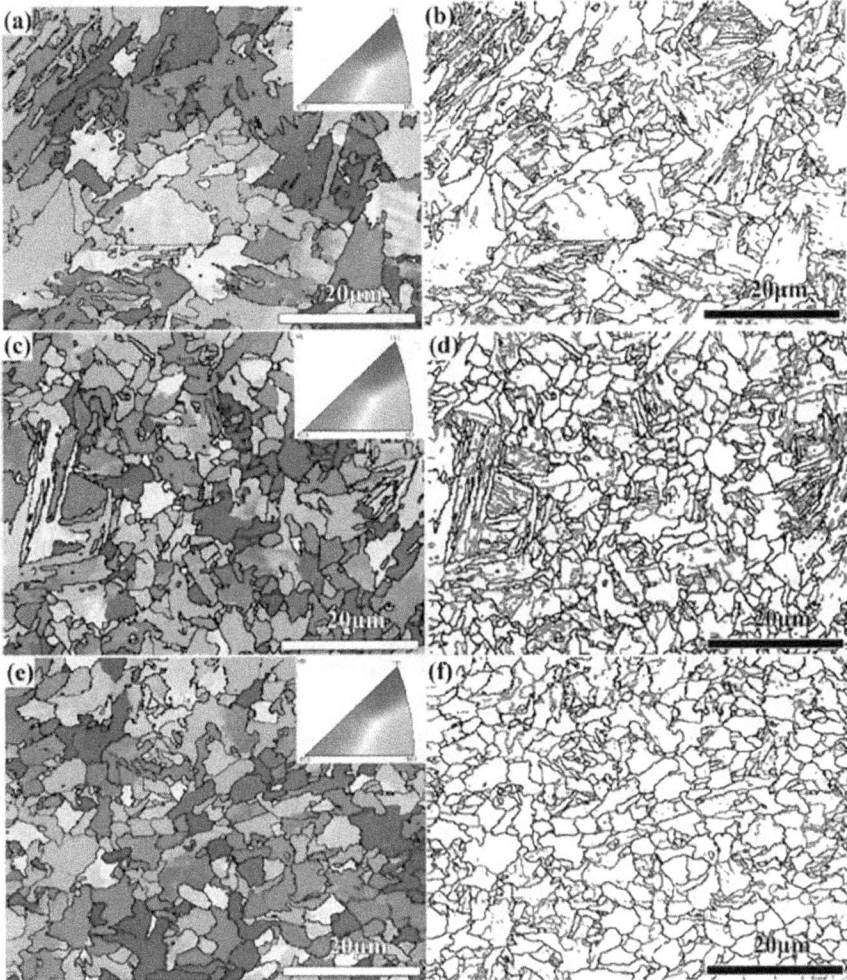

FIGURE 3.6 Different laser powers were utilized in the creation of the grain boundaries and EBSD maps for the samples [129].

through the utilization of the energy input from the reaction between Ti and B. To accomplish this goal, an exothermic reaction was carried out using the energy gained from the reaction between titanium and boron [1]. Because of the impact it has on the rate at which the final product solidifies, the enthalpy of mixing has the ability to cause changes in the microstructure and phases of the product as a whole. They believe that a highly localized heat of reaction is responsible for the influence that it has on the rate of solidification, which they assume is related to the temperature differential that exists. In other words, they believe that the influence that it has on the rate of solidification is caused by a highly localized heat of reaction.

In the LDED process, alloys that are composed of elements that experience an exothermic reaction have a greater chance of developing the kind of microstructures that are the result of rapid solidification. This is because an exothermic reaction generates more heat than a reaction that does not generate heat. On the other hand, this goes against the common knowledge, which suggests that the rate of solidification will quicken if the melt superheat is increased. This interpretation of the data contradicts what should be a straightforward relationship between the two variables [131]. In addition, it was demonstrated that by utilizing the LDED and LPBF approaches, one could bring about a reduction in the melt cooling rate by inducing a rise in the receiving energy density. This was done by proving that this would be the case. This has been demonstrated to be the case. The microstructures that were presented were made by utilizing two different element mixes, the LDED process, and continuing with the procedure in the same way for all of the subsequent steps. Researchers studying a uniform structure in a Ti-Cr alloy noted the alloy's negative heat of reaction, which facilitates the amalgamating method. There was a mention of this in this chapter. The Ti-Nb alloy exhibited abrupt changes in chemical composition between successive build coats [132]. These fluctuations were induced by the fact that the alloy was built in layers. The presence of Nb particles that had not melted was demonstrated by these alterations, which were seen in the voids between the layers.

Surface tension, also called the "Marangoni effect", is assumed to be the primary driver of fluid movement in the melt pool as a direct outcome of the AM process. Several investigations lend credence to this notion. The Marangoni convection that takes place in the melt pool is essential to the process of mass transfer that takes place there, and it also serves to promote element diffusion [133]. Because of this, it has the potential to be a deterministic component in AM techniques for the synthesis of in situ alloys, and this opens up a lot of doors for prospective applications. As a general rule, the water temperature will be higher in the pool's centre than at its solid/liquid boundary. The area in the pool's equatorial centre is affected by the heat source. This is because the energy source is located in a relatively small area, which contributes a considerable amount to the overall heating of the melt pool in its immediate vicinity. The top surface would, as a direct consequence of this, exhibit a gradient in surface tension, which would increase the flow of local melt. This would result in the melting process being accelerated. The Marangoni force causes the molten material to migrate away from the centre of the pool and toward its perimeter, resulting in the formation of a convection pattern. This force is only effective for metals and alloys that do not have an active surface element that modifies the surface tension [134]. Because of this, the material that is located on the surface will migrate lower, and the portion of the melt that is now located at the bottom will be elevated upward. The Marangoni force, in addition to assisting the diffusion process, pulls particles into the melt pool, where they completely melt, leading to chemical uniformity in the AM product that is generated in situ. This occurs because the particles are dragged into the melt pool by the Marangoni force. The Marangoni effect is the consequence of this. A change in the gradient of the liquid's surface tension that happens as a result of the chemical composition of the liquid may also have an influence on the pool geometry. This change can occur as a result of the chemical composition of the liquid. On the other hand, the chemical composition of the alloy can be adjusted to more closely match the dominant texture that was sought. This is possible despite the fact that the geometry of the melt pool is the fundamental component that dictates its mode of solidification [135]. This is accomplished by taking advantage of the fact that the Marangoni force increases as the input energy increases.

3.7   OBSTACLES AND PROSPECTIVE

This trend may be seen in both academic and commercial contexts, and it is largely attributable to recent advancements in the technology that underpins AM of metals. The growth of applications for AM of metals is occurring at a rapid rate. As a consequence of this, the primary focus of this review was directed toward the most recent discoveries made in the body of previously published research concerning the conceptual framework underlying the in situ alloying process as well as the practical applications of this method for the production of AM materials. At this point in the production procedure, the feedstock material can be produced using either powder that has been elementally blended or powder that has been pre-alloyed. Both of these powders can be purchased from a supplier. In addition to the value enhancements that are an inevitable by-product of the characteristics of the AM processes, both types of feedstock powders provide a variety of other advantages that cannot be compared to one another [123]. Because traditional pre-alloyed powders are now easier to obtain, there has been an uptick in demand for these powders, which may be at least partially attributable to this development.

However, the cost of using AM to create components from these new materials, beginning with pre-alloyed powder, puts them out of reach for the vast majority of scientific endeavours. In contrast, in situ alloying is a promising method for rapidly and affordably developing novel alloy compositions for AM. In spite of the challenges inherent to the method, in situ alloying offers several ways in which the materials production process for AM could be enhanced. These properties are important both on their own and when combined with those of other elements. When utilizing this strategy, however, they also constitute one of the most significant obstacles that can be found in the construction of alloys [136].

In addition, the inhomogeneity of the components, as well as any elemental segregation that may take place, has the potential to have a negative effect on the qualities of the finished product. To put it another way, the separation of elements results in a subtle shift in the alloy's local chemical makeup. Because of this, different parts will each have their own unique combination of mechanical and physical qualities, which will make the operations of post-treatment significantly more difficult [137]. In addition, chemical segregation has the ability to change the sample's susceptibility to cracking, which has an effect on the sample's mechanical qualities and needs to be examined on an individual basis. This part of the mechanical qualities of the sample must be looked at as a whole.

Liquid enrichment of Al, Ti, and Ta may cause local tensile stresses and cracking in Ni-based superalloys, and Zr is said to exacerbate the development of solidification cracks in the LPBF of IN738LC. The likelihood of fracture formation in the LPBF of IN738LC increases when Zr is present, which supports both of these claims. The evidence supports both of these assertions. However, Mo and W segregation along the contact in Ti-based alloys may improve thermal characteristics by limiting the growth of microvoids in the interface. Distinctions are made along the contact to achieve this goal. It is possible that the supplementary characteristics will cause heterogeneous nuclei to form during solidification. This will allow the columnar grain shape to be transformed into an equiaxed grain shape [137–140]. The propensity of AM products to develop cracks can also be affected by adjusting the solidification range of the alloy. There are still a number of knowledge gaps that need to be filled in through the completion of additional research activities as the development of this design paradigm moves forward.

3.8   CONCLUSIONS AND SUMMARY

This book gives an overview of the experiments that have been carried out on the temperature and thermal phenomena that occur during laser-based AM. These investigations have been undertaken on a variety of different materials. In addition to providing an introduction of common monitoring devices and thermal models, the presenter also gave a presentation on the connection between the parameters of the process, the thermal behaviour of the product, and the quality of the product. In conjunction with another presentation that was delivered before, this presentation was provided. The production of samples through AM in situ alloying, which requires the use of pure elemental mixes as different to pre-alloyed powders, is attracting an increasing amount of interest from an increasing number of AM firms. This is due to the fact that this method requires the use of pure elemental mixes. The LDED method was utilized in an overwhelming majority of the works that made use of laser-based procedures. In spite of this, the number of works that employ LPBF techniques has greatly expanded, as has the variety of these works, according to the statistics. It is anticipated that the LPBF method would supplant the LDED methodology as the gold standard of the industry within the next several years. In spite of the fact that the LDED method appears to have certain advantages for in situ alloying, this outcome has been observed. One theory that could account for this is that there has been a recent spike in the number of LPBF machines that are made available in research facilities and laboratories. When beginning with a combination of pure elements, it is vital to take into consideration a wide variety of qualities that are shared across the elements. The size of the element, its melting temperature, its reflectivity, its viscosity, its density, or its thermal conductivity could be examples of these disparities. To begin with and most importantly, one must have an understanding of the process phenomena in order to manage the adaptation of proportional elements. When working with components that have not been tampered with in any way, one of the most important considerations is to make certain that the finished product is chemically consistent. This evaluation has the potential to assist in the resolution of quality issues given that the reader has an awareness of the thermal behaviour. However, in order for this to happen, the reader must first have a comprehension of the thermal behaviour.

REFERENCES

1. Simons, M.: Additive manufacturing-a revolution in progress? Insights from a multiple case study. *Int. J. Adv. Manuf. Technol.* 96, 735–749 (2018). https://doi.org/10.1007/s00170-018-1601-1.
2. Prashar, G., Vasudev, H., Bhuddhi, D.: Additive manufacturing: expanding 3D printing horizon in industry 4.0. *Int. J. Interact. Des. Manuf.* (2022). https://doi.org/10.1007/s12008-022-00956-4.
3. Prashar, G., Vasudev, H.: A comprehensive review on sustainable cold spray additive manufacturing: State of the art, challenges and future challenges. *J. Clean. Prod.* 310, 127606 (2021). https://doi.org/10.1016/j.jclepro.2021.127606.
4. Singh, R., Toseef, M., Kumar, J., Singh, J.: Benefits and challenges in additive manufacturing and its applications. In: Kaushal, S., Singh, I., Singh, S., and Gupta, A. (eds.) *Sustainable Advanced Manufacturing and Materials Processing.* pp. 137–157. CRC Press, Boca Raton (2022). https://doi.org/10.1201/9781003269298-8.
5. Tepylo, N., Huang, X., Patnaik, P.C.: Laser-based additive manufacturing technologies for aerospace applications. *Adv. Eng. Mater.* 21, 1900617 (2019). https://doi.org/10.1002/adem.201900617.
6. Singh, J.: A review on mechanisms and testing of wear in slurry pumps, pipeline circuits and hydraulic turbines. *J. Tribol.* 143, 1–83 (2021). https://doi.org/10.1115/1.4050977.
7. Singh, J., Singh, S.: Neural network prediction of slurry erosion of heavy-duty pump impeller/casing materials 18Cr-8Ni, 16Cr-10Ni-2Mo, super duplex 24Cr-6Ni-3Mo-N, and grey cast iron. *Wear.* 476, 203741 (2021). https://doi.org/10.1016/j.wear.2021.203741.
8. Singh, J., Kumar, S., Mohapatra, S.K.: An erosion and corrosion study on thermally sprayed WC-Co-Cr powder synergized with $Mo_2C/Y_2O_3/ZrO_2$ feedstock powders. *Wear.* 438–439, 102751 (2019). https://doi.org/10.1016/j.wear.2019.01.082.
9. Singh, J., Kumar, S., Mohapatra, S.K.: Tribological analysis of WC–10Co–4Cr and Ni–$20Cr_2O_3$ coating on stainless steel 304. *Wear.* 376–377, 1105–1111 (2017). https://doi.org/10.1016/j.wear.2017.01.032.
10. Singh, J.: Slurry erosion performance analysis and characterization of high-velocity oxy-fuel sprayed Ni and Co hardsurfacing alloy coatings. *J. King Saud Univ. - Eng. Sci.* (2021). https://doi.org/10.1016/j.jksues.2021.06.009.
11. Singh, J.: Tribo-performance analysis of HVOF sprayed 86WC-10Co4Cr & Ni-Cr_2O_3 on AISI 316L steel using DOE-ANN methodology. *Ind. Lubr. Tribol.* 73, 727–735 (2021). https://doi.org/10.1108/ILT-04-2020-0147.
12. Singh, J., Singh, S.: Neural network supported study on erosive wear performance analysis of Y_2O_3/WC-10Co4Cr HVOF coating. *J. King Saud Univ. - Eng. Sci.* (2022). https://doi.org/10.1016/j.jksues.2021.12.005.
13. Singh, J.: Wear performance analysis and characterization of HVOF deposited Ni–$20Cr_2O_3$, Ni–$30Al_2O_3$, and Al_2O_3–$13TiO_2$ coatings. *Appl. Surf. Sci. Adv.* 6, 100161 (2021). https://doi.org/10.1016/j.apsadv.2021.100161.
14. Singh, J., Kumar, S., Mohapatra, S.K.: Optimization of erosion wear influencing parameters of HVOF sprayed pumping material for coal-water slurry. *Mater. Today Proc.* 5, 23789–23795 (2018). https://doi.org/10.1016/j.matpr.2018.10.170.
15. Singh, J., Kumar, S., Mohapatra, S.K.: Erosion wear performance of Ni-Cr-O and NiCrBSiFe-WC(Co) composite coatings deposited by HVOF technique. *Ind. Lubr. Tribol.* 71, 610–619 (2019). https://doi.org/10.1108/ILT-04-2018-0149.
16. Singh, J., Kumar, S., Mohapatra, S.K.: Tribological performance of Yttrium (III) and Zirconium (IV) ceramics reinforced WC–10Co4Cr cermet powder HVOF thermally sprayed on X2CrNiMo-17-12-2 steel. *Ceram. Int.* 45, 23126–23142 (2019). https://doi.org/10.1016/j.ceramint.2019.08.007.

17. Singh, J., Kumar, S., Mohapatra, S.K.: Erosion tribo-performance of HVOF deposited Stellite-6 and Colmonoy-88 micron layers on SS-316L. *Tribol. Int.* 147, 105262 (2020). https://doi.org/10.1016/j.triboint.2018.06.004.

18. Singh, J., Mohapatra, S.K., Kumar, S.: Performance analysis of pump materials employed in bottom ash slurry erosion conditions. *J. Tribol.* 30, 73–89 (2021).

19. Singh, J.: Analysis on suitability of HVOF sprayed Ni-20Al, Ni-20Cr and Al-20Ti coatings in coal-ash slurry conditions using artificial neural network model. *Ind. Lubr. Tribol.* 71, 972–982 (2019). https://doi.org/10.1108/ILT-12-2018-0460.

20. Sharma, Y., Singh, K.J., Vasudev, H.: Experimental studies on friction stir welding of aluminium alloys. *Mater. Today Proc.* 50, 2387–2391 (2022). https://doi.org/10.1016/j.matpr.2021.10.254.

21. Satyavathi Yedida, V.V., Vasudev, H.: A review on the development of thermal barrier coatings by using thermal spray techniques. *Mater. Today Proc.* 50, 1458–1464 (2022). https://doi.org/10.1016/j.matpr.2021.09.018.

22. Prashar, G., Vasudev, H.: Surface topology analysis of plasma sprayed Inconel625-Al_2O_3 composite coating. *Mater. Today Proc.* 50, 607–611 (2022). https://doi.org/10.1016/j.matpr.2021.03.090.

23. Sunitha, K., Vasudev, H.: A short note on the various thermal spray coating processes and effect of post-treatment on Ni-based coatings. *Mater. Today Proc.* 50, 1452–1457 (2022). https://doi.org/10.1016/j.matpr.2021.09.017.

24. Ganesh Reddy Majji, B., Vasudev, H., Bansal, A.: A review on the oxidation and wear behavior of the thermally sprayed high-entropy alloys. *Mater. Today Proc.* 50, 1447–1451 (2022). https://doi.org/10.1016/j.matpr.2021.09.016.

25. Parkash, J., Saggu, H.S., Vasudev, H.: A short review on the performance of high velocity oxy-fuel coatings in boiler steel applications. *Mater. Today Proc.* 50, 1442–1446 (2022). https://doi.org/10.1016/j.matpr.2021.09.014.

26. Singh, J., Vasudev, H., Singh, S.: Performance of different coating materials against high temperature oxidation in boiler tubes – A review. *Mater. Today Proc.* 26, 972–978 (2020). https://doi.org/10.1016/j.matpr.2020.01.156.

27. Mehta, A., Vasudev, H., Singh, S.: Recent developments in the designing of deposition of thermal barrier coatings – A review. *Mater. Today Proc.* 26, 1336–1342 (2020). https://doi.org/10.1016/j.matpr.2020.02.271.

28. Singh, J., Singh, J.P.: Numerical analysis on solid particle erosion in elbow of a slurry conveying circuit. *J. Pipeline Syst. Eng. Pract.* 12, 04020070 (2021). https://doi.org/10.1061/(asce)ps.1949-1204.0000518.

29. Sunitha, K. and Vasudev, H.: Microstructural and mechanical characterization of HVOF-sprayed Ni-based alloy coating. *Int. J. Surf. Eng. Interdiscip. Mater. Sci.* 10, 1–9 (2022). https://doi.org/10.4018/IJSEIMS.298705.

30. Prashar, G., Vasudev, H.: Hot corrosion behavior of super alloys. *Mater. Today Proc.* 26, 1131–1135 (2020). https://doi.org/10.1016/j.matpr.2020.02.226.

31. Singh, M., Vasudev, H., Kumar, R.: Microstructural characterization of BN thin films using RF magnetron sputtering method. *Mater. Today Proc.* 26, 2277–2282 (2020). https://doi.org/10.1016/j.matpr.2020.02.493.

32. Singh, J.: Application of thermal spray coatings for protection against erosion, abrasion, and corrosion in hydropower plants and offshore industry. In: Thakur, L. and Vasudev, H. (eds.) *Thermal Spray Coatings.* pp. 243–283. CRC Press, Boca Raton (2021). https://doi.org/10.1201/9781003213185-10.

33. Singh, J., Singh, S., Gill, R.: Applications of biopolymer coatings in biomedical engineering. *J. Electrochem. Sci. Eng.* (2022). https://doi.org/10.5599/jese.1460.

34. Singh, J., Kumar, S., Singh, J.P., Kumar, P., Mohapatra, S.K.: CFD modeling of erosion wear in pipe bend for the flow of bottom ash suspension. *Part. Sci. Technol.* 37, 275–285 (2019). https://doi.org/10.1080/02726351.2017.1364816.

35. Singh, J., Singh, S., Pal Singh, J.: Investigation on wall thickness reduction of hydro-power pipeline underwent to erosion-corrosion process. *Eng. Fail. Anal.* 127, 105504 (2021). https://doi.org/10.1016/j.engfailanal.2021.105504.

36. Singh, J., Kumar, S., Singh, G.: Taguchi's approach for optimization of tribo-resistance parameters Forss304. *Mater. Today Proc.* 5, 5031–5038 (2018). https://doi.org/10.1016/j.matpr.2017.12.081.

37. Singh, J., Singh, J.P.: Performance analysis of erosion resistant Mo_2C reinforced WC-CoCr coating for pump impeller with Taguchi's method. *Ind. Lubr. Tribol.* 74, 431–441 (2022). https://doi.org/10.1108/ILT-05-2020-0155.

38. Kumar, D., Yadav, R., Singh, J.: Evolution and adoption of microwave claddings in modern engineering applications. In: *Advances in Microwave Processing for Engineering Materials.* pp. 134–153. CRC Press, Boca Raton (2022). https://doi.org/10.1201/9781003248743-8.

39. Singh, J., Gill, H.S., Vasudev, H.: Computational fluid dynamics analysis on effect of particulate properties on erosive degradation of pipe bends. *Int. J. Interact. Des. Manuf.* (2022). https://doi.org/10.1007/s12008-022-01094-7.

40. Singh, J., Singh, R., Singh, S., Vasudev, H., Kumar, S.: Reducing scrap due to missed operations and machining defects in 90PS pistons. *Int. J. Interact. Des. Manuf.* (2022). https://doi.org/10.1007/s12008-022-01071-0.

41. Singh, J., Kumar, S., Mohapatra, S.K.: Study on solid particle erosion of pump materials by fly ash slurry using Taguchi's orthogonal array. *Tribol. - Finnish J. Tribol.* 38, 31–38 (2021). https://doi.org/10.30678/fjt.97530.

42. Singh, J., Singh, J.P., Singh, M., Szala, M.: Computational analysis of solid particle-erosion produced by bottom ash slurry in 90° elbow. In *MATEC Web of Conferences,* vol. 252, p. 04008 (2019). https://doi.org/0.1051/matecconf/201925204008.

43. Singh, J., Kumar, S., Mohapatra, S.: Study on role of particle shape in erosion wear of austenitic steel using image processing analysis technique. *Proc. Inst. Mech. Eng. Part J J. Eng. Tribol.* 233, 712–725 (2019). https://doi.org/10.1177/1350650118794698.

44. Singh, J.: *Investigation on Slurry Erosion of Different Pumping Materials and Coatings.* Thapar Institute of Engineering and Technology, Patiala, India (2019).

45. Vasudev, H., Thakur, L., Singh, H., Bansal, A.: Mechanical and microstructural behaviour of wear resistant coatings on cast iron lathe machine beds and slides. *Met. Mater.* 56, 55–63 (2018). https://doi.org/10.4149/km_2018_1_55.

46. Vasudev, H., Singh, G., Bansal, A., Vardhan, S., Thakur, L.: Microwave heating and its applications in surface engineering: a review. *Mater. Res. Express.* 6, 102001 (2019). https://doi.org/10.1088/2053-1591/ab3674.

47. Vasudev, H., Thakur, L., Bansal, A., Singh, H., Zafar, S.: High temperature oxidation and erosion behaviour of HVOF sprayed bi-layer Alloy-718/NiCrAlY coating. *Surf. Coatings Technol.* 362, 366–380 (2019). https://doi.org/10.1016/j.surfcoat.2019.02.012.

48. Bansal, A., Vasudev, H., Sharma, A.K., Kumar, P.: Investigation on the effect of post weld heat treatment on microwave joining of the Alloy-718 weldment. *Mater. Res. Express.* 6, 086554 (2019). https://doi.org/10.1088/2053-1591/ab1d9a.

49. Vasudev, H., Singh, P., Thakur, L., Bansal, A.: Mechanical and microstructural characterization of microwave post processed Alloy-718 coating. *Mater. Res. Express.* 6, 1265f5 (2020). https://doi.org/10.1088/2053-1591/ab66fb.

50. Vasudev, H.: Wear characteristics of Ni-WC powder deposited by using a microwave route on mild steel. *Int. J. Surf. Eng. Interdiscip. Mater. Sci.* 8, 44–54 (2020). https://doi.org/10.4018/IJSEIMS.2020010104.

51. Singh, G., Vasudev, H., Bansal, A., Vardhan, S., Sharma, S.: Microwave cladding of Inconel-625 on mild steel substrate for corrosion protection. *Mater. Res. Express.* 7, 026512 (2020). https://doi.org/10.1088/2053-1591/ab6fa3.

52. Vasudev, H., Thakur, L., Singh, H., Bansal, A.: A study on processing and hot corrosion behaviour of HVOF sprayed Inconel718-nano Al₂O₃ coatings. *Mater. Today Commun.* 25, 101626 (2020). https://doi.org/10.1016/j.mtcomm.2020.101626.
53. Prashar, G., Vasudev, H., Thakur, L.: Performance of different coating materials against slurry erosion failure in hydrodynamic turbines: A review. *Eng. Fail. Anal.* 115, 104622 (2020). https://doi.org/10.1016/j.engfailanal.2020.104622.
54. Vasudev, H., Thakur, L., Singh, H., Bansal, A.: An investigation on oxidation behaviour of high velocity oxy-fuel sprayed Inconel718-Al₂O₃ composite coatings. *Surf. Coatings Technol.* 393, 125770 (2020). https://doi.org/10.1016/j.surfcoat.2020.125770.
55. Prashar, G., & Vasudev, H.: Understanding cold spray technology for hydroxyapatite deposition: Review paper. *J. Electrochem. Sci. Eng.* 13, 41–62 (2023). https://doi.org/10.5599/jese.1424.
56. Prashar, G., Vasudev, H., Thakur, L.: Influence of heat treatment on surface properties of HVOF deposited WC and Ni-based powder coatings: a review. *Surf. Topogr. Metrol. Prop.* 9, 043002 (2021). https://doi.org/10.1088/2051-672X/ac3a52.
57. Vasudev, H., Prashar, G., Thakur, L., Bansal, A.: Electrochemical corrosion behavior and microstructural characterization of HVOF sprayed Inconel-718 coating on gray cast iron. *J. Fail. Anal. Prev.* 21, 250–260 (2021). https://doi.org/10.1007/s11668-020-01057-8.
58. Singh, M., Vasudev, H., Kumar, R.: Corrosion and tribological behaviour of BN thin films deposited using magnetron sputtering. *Int. J. Surf. Eng. Interdiscip. Mater. Sci.* 9, 24–39 (2021). https://doi.org/10.4018/IJSEIMS.2021070102.
59. Singh, P., Bansal, A., Vasudev, H., Singh, P.: In situ surface modification of stainless steel with hydroxyapatite using microwave heating. *Surf. Topogr. Metrol. Prop.* 9, 035053 (2021). https://doi.org/10.1088/2051-672X/ac28a9.
60. Prashar, G., Vasudev, H.: High temperature erosion behavior of plasma sprayed Al₂O₃ coating on AISI-304 stainless steel. *World J. Eng.* 18, 760–766 (2021). https://doi.org/10.1108/WJE-10-2020-0476.
61. Vasudev, H., Prashar, G., Thakur, L., Bansal, A.: Microstructural characterization and electrochemical corrosion behaviour of HVOF sprayed Alloy718-nanoAl₂O₃ composite coatings. *Surf. Topogr. Metrol. Prop.* 9, 035003 (2021). https://doi.org/10.1088/2051-672X/ac1044.
62. Vasudev, H., Thakur, L., Singh, H., Bansal, A.: Erosion behaviour of HVOF sprayed Alloy718-nano Al₂O₃ composite coatings on grey cast iron at elevated temperature conditions. *Surf. Topogr. Metrol. Prop.* 9, 035022 (2021). https://doi.org/10.1088/2051-672X/ac1c80.
63. Singh, G., Vasudev, H., Bansal, A., Vardhan, S.: Influence of heat treatment on the microstructure and corrosion properties of the Inconel-625 clad deposited by microwave heating. *Surf. Topogr. Metrol. Prop.* 9, 025019 (2021). https://doi.org/10.1088/2051-672X/abfc61.
64. Prashar, G., Vasudev, H.: Structure-property correlation and high-temperature erosion performance of Inconel625-Al₂O₃ plasma-sprayed bimodal composite coatings. *Surf. Coatings Technol.* 439, 128450 (2022). https://doi.org/10.1016/j.surfcoat.2022.128450.
65. Prashar, G., Vasudev, H.: Structure–property correlation of plasma-sprayed Inconel625-Al₂O₃ bimodal composite coatings for high-temperature oxidation protection. *J. Therm. Spray Technol.* 31, 2385–2408 (2022). https://doi.org/10.1007/s11666-022-01466-1.
66. Singh, M., Vasudev, H., Singh, M.: Surface protection of SS-316L with boron nitride based thin films using radio frequency magnetron sputtering technique. *J. Electrochem. Sci. Eng.* 12, 851–863 (2022). https://doi.org/10.5599/jese.1247.
67. Vasudev, H., Thakur, L., Singh, H., Bansal, A.: Effect of addition of Al₂O₃ on the high-temperature solid particle erosion behaviour of HVOF sprayed Inconel-718 coatings. *Mater. Today Commun.* 30, 103017 (2022). https://doi.org/10.1016/j.mtcomm.2021.103017.

68. Vasudev, H., Prashar, G., Thakur, L., Bansal, A.: Electrochemical corrosion behavior and microstructural characterization of HVOF sprayed Inconel718-Al$_2$O$_3$ composite coatings. *Surf. Rev. Lett.* 29, 2250017 (2022). https://doi.org/10.1142/S0218625X22500172.

69. Singh, P., Vasudev, H., Bansal, A.: Effect of post-heat treatment on the microstructural, mechanical, and bioactivity behavior of the microwave-assisted alumina-reinforced hydroxyapatite cladding. *Proc. Inst. Mech. Eng. Part E J. Process Mech. Eng.* 095440892211161 (2022). https://doi.org/10.1177/09544089221116168.

70. Prashar, G., Vasudev, H.: A review on the influence of process parameters and heat treatment on the corrosion performance of Ni-based thermal spray coatings. *Surf. Rev. Lett.* 29, 1–18 (2022). https://doi.org/10.1142/S0218625X22300015.

71. Prashar, G., Vasudev, H., Thakur, L.: High-temperature oxidation and erosion resistance of Ni-based thermally-sprayed coatings used in power generation machinery: A review. *Surf. Rev. Lett.* 29, 2230003 (2022). https://doi.org/10.1142/S0218625X22300039.

72. Singh, J., Singh, S., Verma, A.: Artificial intelligence in use of ZrO$_2$ material in biomedical science. *J. Electrochem. Sci. Eng.* (2022). https://doi.org/10.5599/jese.1498.

73. Singh, J., Singh, S.: A review on machine learning aspect in physics and mechanics of glasses. *Mater. Sci. Eng. B.* 284, 115858 (2022). https://doi.org/10.1016/j.mseb.2022.115858.

74. Sahasrabudhe, H., Bandyopadhyay, A.: Laser-based additive manufacturing of zirconium. *Appl. Sci.* 8, 393 (2018). https://doi.org/10.3390/app8030393.

75. Schmidt, M., Merklein, M., Bourell, D., Dimitrov, D., Hausotte, T., Wegener, K., Overmeyer, L., Vollertsen, F., Levy, G.N.: Laser based additive manufacturing in industry and academia. *CIRP Ann.* 66, 561–583 (2017). https://doi.org/10.1016/j.cirp.2017.05.011.

76. Romano, J., Ladani, L., Sadowski, M.: Thermal modeling of laser based additive manufacturing processes within common materials. *Procedia Manuf.* 1, 238–250 (2015). https://doi.org/10.1016/j.promfg.2015.09.012.

77. Hu, D., Kovacevic, R.: Sensing, modeling and control for laser-based additive manufacturing. *Int. J. Mach. Tools Manuf.* 43, 51–60 (2003). https://doi.org/10.1016/S0890-6955(02)00163-3.

78. Prakash, C., Uddin, M.S.: Surface modification of β-phase Ti implant by hydroaxyapatite mixed electric discharge machining to enhance the corrosion resistance and in-vitro bioactivity. *Surf. Coatings Technol.* 326, 134–145 (2017). https://doi.org/10.1016/j.surfcoat.2017.07.040.

79. Pradhan, S., Singh, S., Prakash, C., Królczyk, G., Pramanik, A., Pruncu, C.I.: Investigation of machining characteristics of hard-to-machine Ti-6Al-4V-ELI alloy for biomedical applications. *J. Mater. Res. Technol.* 8, 4849–4862 (2019). https://doi.org/10.1016/j.jmrt.2019.08.033.

80. Prakash, C., Kansal, H.K., Pabla, B., Puri, S., Aggarwal, A.: Electric discharge machining – A potential choice for surface modification of metallic implants for orthopedic applications: A review. *Proc. Inst. Mech. Eng. Part B J. Eng. Manuf.* 230, 331–353 (2016). https://doi.org/10.1177/0954405515579113.

81. Prakash, C., Singh, S., Singh, M., Verma, K., Chaudhary, B., Singh, S.: Multi-objective particle swarm optimization of EDM parameters to deposit HA-coating on biodegradable Mg-alloy. *Vacuum.* 158, 180–190 (2018). https://doi.org/10.1016/j.vacuum.2018.09.050.

82. Prakash, C., Kansal, H.K., Pabla, B.S., Puri, S.: Powder mixed electric discharge machining: An innovative surface modification technique to enhance fatigue performance and bioactivity of β-Ti implant for orthopedics application. *J. Comput. Inf. Sci. Eng.* 16, (2016). https://doi.org/10.1115/1.4033901.

83. Prakash, C., Singh, S., Pabla, B.S., Uddin, M.S.: Synthesis, characterization, corrosion and bioactivity investigation of nano-HA coating deposited on biodegradable Mg-Zn-Mn alloy. *Surf. Coatings Technol.* 346, 9–18 (2018). https://doi.org/10.1016/j.surfcoat.2018.04.035.

84. Prakash, C., Singh, S., Pruncu, C., Mishra, V., Królczyk, G., Pimenov, D., Pramanik, A.: Surface modification of Ti-6Al-4V alloy by electrical discharge coating process using partially sintered Ti-Nb electrode. *Materials*. 12, 1006 (2019). https://doi.org/10.3390/ma12071006.

85. Prakash, C., Kansal, H.K., Pabla, B.S., Puri, S.: Multi-objective optimization of powder mixed electric discharge machining parameters for fabrication of biocompatible layer on β-Ti alloy using NSGA-II coupled with Taguchi based response surface methodology. *J. Mech. Sci. Technol*. 30, 4195–4204 (2016). https://doi.org/10.1007/s12206-016-0831-0.

86. Prakash, C., Kansal, H.K., Pabla, B.S., Puri, S.: Processing and characterization of novel biomimetic nanoporous bioceramic surface on β-Ti implant by powder mixed electric discharge machining. *J. Mater. Eng. Perform*. 24, 3622–3633 (2015). https://doi.org/10.1007/s11665-015-1619-6.

87. Prakash, C., Kansal, H.K., Pabla, B.S., Puri, S.: Experimental investigations in powder mixed electric discharge machining of Ti–35Nb–7Ta–5Zrβ-titanium alloy. *Mater. Manuf. Process*. 32, 274–285 (2017). https://doi.org/10.1080/10426914.2016.1198018.

88. Nair, A., Ramkumar, P., Mahadevan, S., Prakash, C., Dixit, S., Murali, G., Vatin, N.I., Epifantsev, K., Kumar, K.: Machine learning for prediction of heat pipe effectiveness. *Energies*. 15, 3276 (2022). https://doi.org/10.3390/en15093276.

89. Tiwari, A.K., Kumar, A., Kumar, N., Prakash, C.: Investigation on micro-residual stress distribution near hole using nanoindentation: Effect of drilling speed. *Meas. Control*. 52, 1252–1263 (2019). https://doi.org/10.1177/0020294019858107.

90. Pramanik, A., Basak, A.K., Littlefair, G., Debnath, S., Prakash, C., Singh, M.A., Marla, D., Singh, R.K.: Methods and variables in electrical discharge machining of titanium alloy – A review. *Heliyon*. 6, e05554 (2020). https://doi.org/10.1016/j.heliyon.2020.e05554.

91. Chowdhury, S., Yadaiah, N., Prakash, C., Ramakrishna, S., Dixit, S., Raj Gupta, L., Buddhi, D.,: Laser powder bed fusion: a state-of-the-art review of the technology, materials, properties & defects, and numerical modelling. *Journal of Materials Research and Technology*. 20, 2109–2172 (2022). https://doi.org/10.1016/j.jmrt.2022.07.121.

92. Zheng Yang, K., Pramanik, A., Basak, A.K., Dong, Y., Prakash, C., Shankar, S., Dixit, S., Kumar, K., Ivanovich Vatin, N.: Application of coolants during tool-based machining – A review. *Ain Shams Eng. J*. (2022). https://doi.org/10.1016/j.asej.2022.101830.

93. Usca, Ü.A., Uzun, M., Şap, S., Giasin, K., Pimenov, D.Y., Prakash, C.: Determination of machinability metrics of AISI 5140 steel for gear manufacturing using different cooling/lubrication conditions. *J. Mater. Res. Technol*. (2022). https://doi.org/10.1016/j.jmrt.2022.09.067.

94. Nguyen, D.N., Le Chau, N., Dao, T.P., Prakash, C., Singh, S.: Experimental study on polishing process of cylindrical roller bearings. *Meas. Control*. 52, 1272–1281 (2019). https://doi.org/10.1177/0020294019864395.

95. Sreenivasan, R., Goel, A., Bourell, D.L.: Sustainability issues in laser-based additive manufacturing. *Phys. Procedia*. 5, 81–90 (2010). https://doi.org/10.1016/j.phpro.2010.08.124.

96. Francis, J., Bian, L.: Deep learning for distortion prediction in laser-based additive manufacturing using big data. *Manuf. Lett*. 20, 10–14 (2019). https://doi.org/10.1016/j.mfglet.2019.02.001.

97. Thompson, S.M., Bian, L., Shamsaei, N., Yadollahi, A.: An overview of direct laser deposition for additive manufacturing; part I: Transport phenomena, modeling and diagnostics. *Addit. Manuf*. 8, 36–62 (2015). https://doi.org/10.1016/j.addma.2015.07.001.

98. Panda, B.K., Sahoo, S.: Numerical simulation of residual stress in laser based additive manufacturing process. *IOP Conf. Ser. Mater. Sci. Eng*. 338, 012030 (2018). https://doi.org/10.1088/1757-899X/338/1/012030.

99. Shiva, S., Palani, I.A., Mishra, S.K., Paul, C.P., Kukreja, L.M.: Investigations on the influence of composition in the development of Ni–Ti shape memory alloy using laser based additive manufacturing. *Opt. Laser Technol.* 69, 44–51 (2015). https://doi.org/10.1016/j.optlastec.2014.12.014.

100. Wang, C., Tan, X.P., Du, Z., Chandra, S., Sun, Z., Lim, C.W.J., Tor, S.B., Lim, C.S., Wong, C.H.: Additive manufacturing of NiTi shape memory alloys using pre-mixed powders. *J. Mater. Process. Technol.* 271, 152–161 (2019). https://doi.org/10.1016/j.jmatprotec.2019.03.025.

101. Fu, Y., Downey, A.R.J., Yuan, L., Zhang, T., Pratt, A., Balogun, Y.: Machine learning algorithms for defect detection in metal laser-based additive manufacturing: A review. *J. Manuf. Process.* 75, 693–710 (2022). https://doi.org/10.1016/j.jmapro.2021.12.061.

102. Soro, N., Attar, H., Brodie, E., Veidt, M., Molotnikov, A., Dargusch, M.S.: Evaluation of the mechanical compatibility of additively manufactured porous Ti–25Ta alloy for load-bearing implant applications. *J. Mech. Behav. Biomed. Mater.* 97, 149–158 (2019). https://doi.org/10.1016/j.jmbbm.2019.05.019.

103. Seifi, S.H., Tian, W., Doude, H., Tschopp, M.A., Bian, L.: Layer-wise modeling and anomaly detection for laser-based additive manufacturing. *J. Manuf. Sci. Eng.* 141 (2019). https://doi.org/10.1115/1.4043898.

104. Aboutaleb, A.M., Bian, L., Shamsaei, N., Thompson, S.M.: Systematic optimization of Laser-based Additive Manufacturing for multiple mechanical properties. In: *2016 IEEE International Conference on Automation Science and Engineering (CASE)*. pp. 780–785. IEEE (2016). https://doi.org/10.1109/COASE.2016.7743481.

105. Kumar, S., Pityana, S.: Laser-based additive manufacturing of metals. *Adv. Mater. Res.* 227, 92–95 (2011). https://doi.org/10.4028/www.scientific.net/AMR.227.92.

106. Moghaddam, N.S., Jahadakbar, A., Amerinatanzi, A., Elahinia, M.: Recent advances in laser-based additive manufacturing. In: Bian, L., Shamsaei, N., and Usher, J. (eds.) *Laser-Based Additive Manufacturing of Metal Parts*. pp. 1–24. CRC Press, Taylor & Francis, Boca Raton (2017). https://doi.org/10.1201/9781315151441-1.

107. Kundakcioglu, E., Lazoglu, I., Rawal, S.: Transient thermal modeling of laser-based additive manufacturing for 3D freeform structures. *Int. J. Adv. Manuf. Technol.* 85, 493–501 (2016). https://doi.org/10.1007/s00170-015-7932-2.

108. Tian, Q., Guo, S., Melder, E., Bian, L., Guo, W.: Deep learning-based data fusion method for in situ porosity detection in laser-based additive manufacturing. *J. Manuf. Sci. Eng.* 143, (2021). https://doi.org/10.1115/1.4048957.

109. Liu, W.-W., Saleheen, K.M., Tang, Z., Wang, H., Al-Hammadi, G., Abdelrahman, A., Yongxin, Z., Hua, S.-G., Wang, F.-T.: Review on scanning pattern evaluation in laser-based additive manufacturing. *Opt. Eng.* 60, (2021). https://doi.org/10.1117/1.OE.60.7.070901.

110. Priya, P., Mercer, B., Huang, S., Aboukhatwa, M., Yuan, L., Chaudhuri, S.: Towards prediction of microstructure during laser based additive manufacturing process of Co-Cr-Mo powder beds. *Mater. Des.* 196, 109117 (2020). https://doi.org/10.1016/j.matdes.2020.109117.

111. Beese, A.M., Carroll, B.E.: Review of mechanical properties of Ti-6Al-4V made by laser-based additive manufacturing using powder feedstock. *JOM.* 68, 724–734 (2016). https://doi.org/10.1007/s11837-015-1759-z.

112. Jadhav, S.D., Goossens, L.R., Kinds, Y., Van Hooreweder, B., Vanmeensel, K.: Laser-based powder bed fusion additive manufacturing of pure copper. *Addit. Manuf.* 42, 101990 (2021). https://doi.org/10.1016/j.addma.2021.101990.

113. Stathatos, E., Vosniakos, G.-C.: Efficient temperature regulation through power optimization for arbitrary paths in laser based additive manufacturing. *CIRP J. Manuf. Sci. Technol.* 33, 133–142 (2021). https://doi.org/10.1016/j.cirpj.2021.03.008.

114. Jafari-Marandi, R., Khanzadeh, M., Tian, W., Smith, B., Bian, L.: From in-situ monitoring toward high-throughput process control: cost-driven decision-making framework for laser-based additive manufacturing. *J. Manuf. Syst.* 51, 29–41 (2019). https://doi.org/10.1016/j.jmsy.2019.02.005.

115. Bian, L., Shamsaei, N., Usher, J.M. eds: *Laser-Based Additive Manufacturing of Metal Parts.* CRC Press, Taylor & Francis, Boca Raton (2017). https://doi.org/10.1201/9781315151441.

116. Fereiduni, E., Yakout, M., Elbestawi, M.: Laser-based additive manufacturing of lightweight metal matrix composites. In: *Additive Manufacturing of Emerging Materials.* pp. 55–109. Springer International Publishing, Cham (2019). https://doi.org/10.1007/978-3-319-91713-9_3.

117. Gopan, V., Leo Dev Wins, K., Surendran, A.: Innovative potential of additive friction stir deposition among current laser based metal additive manufacturing processes: A review. *CIRP J. Manuf. Sci. Technol.* 32, 228–248 (2021). https://doi.org/10.1016/j.cirpj.2020.12.004.

118. Moesen, M., Craeghs, T., Kruth, J.-P., Schrooten, J.: Robust beam compensation for laser-based additive manufacturing. *Comput. Des.* 43, 876–888 (2011). https://doi.org/10.1016/j.cad.2011.03.004.

119. Khanzadeh, M., Bian, L., Shamsaei, N., Thompson, S.M.: Porosity detection of laser based additive manufacturing using melt pool morphology clustering. In: *Solid Freeform Fabrication 2016: Proceedings of the 26th Annual International Solid Freeform Fabrication Symposium – An Additive Manufacturing Conference Solid Freeform Fabrication 2016: Proceedings of the 27th Annual International.* pp. 1487–1494 (2016).

120. Moylan, S., Whitenton, E., Lane, B., Slotwinski, J.: Infrared thermography for laser-based powder bed fusion additive manufacturing processes. Presented at the (2014). https://doi.org/10.1063/1.4864956.

121. Huang, D., Tan, Q., Zhou, Y., Yin, Y., Wang, F., Wu, T., Yang, X., Fan, Z., Liu, Y., Zhang, J., Huang, H., Yan, M., Zhang, M.-X.: The significant impact of grain refiner on γ-TiAl intermetallic fabricated by laser-based additive manufacturing. *Addit. Manuf.* 46, 102172 (2021). https://doi.org/10.1016/j.addma.2021.102172.

122. Guo, S., Guo, W., Bain, L.: Hierarchical spatial-temporal modeling and monitoring of melt pool evolution in laser-based additive manufacturing. *IISE Trans.* 52, 977–997 (2020). https://doi.org/10.1080/24725854.2019.1704465.

123. Simonelli, M., Aboulkhair, N.T., Cohen, P., Murray, J.W., Clare, A.T., Tuck, C., Hague, R.J.M.: A comparison of Ti-6Al-4V in-situ alloying in selective laser melting using simply-mixed and satellited powder blend feedstocks. *Mater. Charact.* 143, 118–126 (2018). https://doi.org/10.1016/j.matchar.2018.05.039.

124. Wei, C., Li, L.: Recent progress and scientific challenges in multi-material additive manufacturing via laser-based powder bed fusion. *Virtual Phys. Prototyp.* 16, 347–371 (2021). https://doi.org/10.1080/17452759.2021.1928520.

125. Wei, C., Zhang, Z., Cheng, D., Sun, Z., Zhu, M., Li, L.: An overview of laser-based multiple metallic material additive manufacturing: from macro- to micro-scales. *Int. J. Extrem. Manuf.* 3, 012003 (2021). https://doi.org/10.1088/2631-7990/abce04.

126. Gnäupel-Herold, T., Slotwinski, J., Moylan, S.: Neutron measurements of stresses in a test artifact produced by laser-based additive manufacturing. Presented at the (2014). https://doi.org/10.1063/1.4864958.

127. Makineni, S.K., Kini, A.R., Jägle, E.A., Springer, H., Raabe, D., Gault, B.: Synthesis and stabilization of a new phase regime in a Mo-Si-B based alloy by laser-based additive manufacturing. *Acta Mater.* 151, 31–40 (2018). https://doi.org/10.1016/j.actamat.2018.03.037.

128. Noskov, A., Ervik, T.K., Tsivilskiy, I., Gilmutdinov, A., Thomassen, Y.: Characterization of ultrafine particles emitted during laser-based additive manufacturing of metal parts. *Sci. Rep.* 10, 20989 (2020). https://doi.org/10.1038/s41598-020-78073-z.

129. Abiola Raji, S., Patricia Idowu Popoola, A., Leslie Pityana, S., Muhmmed Popoola, O., Olufemi Aramide, F., Tlotleng, M., Kwamina Kum Arthur, N.: Laser based additive manufacturing technology for fabrication of titanium aluminide-based composites in aerospace component applications. In: *Aerodynamics*. IntechOpen (2021). https://doi.org/10.5772/intechopen.85538.

130. Clayton, R.M.: The use of elemental powder mixes in laser-based additive manufacturing (2013). https://scholarsmine.mst.edu/masters_theses/7194.

131. Zhou, Y., Chen, S., Chen, X., Cui, T., Liang, J., Liu, C.: The evolution of bainite and mechanical properties of direct laser deposition 12CrNi2 alloy steel at different laser power. *Mater. Sci. Eng. A.* 742, 150–161 (2019). https://doi.org/10.1016/j.msea.2018.10.092.

132. Paul, C.P., Jinoop, A.N., Kumar, A., Bindra, K.S.: Laser-based metal additive manufacturing: Technology, global scenario and our experiences. *Trans. Indian Natl. Acad. Eng.* 6, 895–908 (2021). https://doi.org/10.1007/s41403-021-00228-9.

133. Khoda, B., Benny, T., Rao, P.K., Sealy, M.P., Zhou, C.: Applications of laser-based additive manufacturing. In: Bian, L., Shamsaei, N., and Usher, J. (eds.) *Laser-Based Additive Manufacturing of Metal Parts.* pp. 239–284. CRC Press, Taylor & Francis, Boca Raton (2017). https://doi.org/10.1201/9781315151441-8.

134. García-Díaz, A., Panadeiro, V., Lodeiro, B., Rodríguez-Araújo, J., Stavridis, J., Papacharalampopoulos, A., Stavropoulos, P.: OpenLMD, an open source middleware and toolkit for laser-based additive manufacturing of large metal parts. *Robot. Comput. Integr. Manuf.* 53, 153–161 (2018). https://doi.org/10.1016/j.rcim.2018.04.006.

135. Chen, Q., Guillemot, G., Gandin, C.-A., Bellet, M.: Numerical modelling of the impact of energy distribution and Marangoni surface tension on track shape in selective laser melting of ceramic material. *Addit. Manuf.* 21, 713–723 (2018). https://doi.org/10.1016/j.addma.2018.03.003.

136. Coleman, J., Plotkowski, A., Stump, B., Raghavan, N., Sabau, A.S., Krane, M.J.M., Heigel, J., Ricker, R.E., Levine, L., Babu, S.S.: Sensitivity of thermal predictions to uncertain surface tension data in laser additive manufacturing. *J. Heat Transfer.* 142 (2020). https://doi.org/10.1115/1.4047916.

137. Mukherjee, T., Manvatkar, V., De, A., DebRoy, T.: Mitigation of thermal distortion during additive manufacturing. *Scr. Mater.* 127, 79–83 (2017). https://doi.org/10.1016/j.scriptamat.2016.09.001.

138. Gribova, V., Kulchin, Y., Nikitin, A., Timchenko, V.: The concept of support for laser-based additive manufacturing on the basis of artificial intelligence methods. Presented at the (2020). https://doi.org/10.1007/978-3-030-59535-7_30.

139. Chen, L., Yao, X., Chew, Y., Weng, F., Moon, S.K., Bi, G.: Data-driven adaptive control for laser-based additive manufacturing with automatic controller tuning. *Appl. Sci.* 10, 7967 (2020). https://doi.org/10.3390/app10227967.

140. Poza, P., Múnez, C.J., Garrido-Maneiro, M.A., Vezzù, S., Rech, S., Trentin, A.: Mechanical properties of Inconel 625 cold-sprayed coatings after laser remelting. Depth sensing indentation analysis. *Surf. Coatings Technol.* 243, 51–57 (2014). https://doi.org/10.1016/j.surfcoat.2012.03.018.

4 Microstructure and Characteristics of Alloys Produced by Additive Laser-Based Manufacturing Technique

Amrinder Mehta, Hitesh Vasudev,
and Chander Prakash
Lovely Professional University

Sharanjit Singh
DAV University

Kuldeep K. Saxena
Lovely Professional University

CONTENTS

DOI: 10.1201/9781003347408-4

4.1   INTRODUCTION

The additive manufacturing (AM) process in metal has seen substantial development during the past 20 years. As a result of the development of metal powder bed AM, it is now possible to build parts with production-quality specifications [1,2]. The ability of AM, which has recently garnered a lot of attention, to generate products with intricate shapes that are either prohibitively expensive or otherwise infeasible to construct using more traditional methods has contributed significantly to this interest. This capacity to produce complicated forms presents a previously unheard-of potential for the oil and gas sector to rethink and optimize a wide range of components, such as cutting heads, heat exchangers, and filtration machinery [3]. This opportunity has the potential to revolutionize the oil and gas industry. The strategy is very different from the traditional processing methods in a number of important respects. These formative and subtractive processes are being phased out in favour of layer-by-layer manufacturing, which is essentially printing with metal. Unprecedented freedom in the manufacturing of difficult structures is made possible by the use of unique materials without the use of additional tooling. This results in extremely high precision and control over the final product [4].

A considerably lower manufacturing lead time is yet another significant advantage that can be derived from AM [5]. This is due to the fact that new designs and components will reach the market more quickly, and consumer demands will be satisfied faster, all while producing far less waste material throughout the manufacturing process. Such benefits, however, required to be evaluated opposed to innate drawbacks, such as the currently higher price of AM equipment, the potential for lengthy build periods, and the potentially expensive and difficult powder feedstocks. A lot of surface engineering techniques are applied to provide the superior surfaces in terms of mechanical, morphological, and physicochemical properties [6–30]. These techniques involve thermal spray coatings [31–47], claddings [48,49], heat treatment [2,49–52], post-treatment [1,53–58], powder bed fusion (PBF) [59–63], etc. The effects of the scanning approach on the microstructure and creep behaviour of Ti_6Al_4V, TiAl, Inconel 625, and cobalt chromium (CoCr) throughout powder bed fusion electron beam additive manufacturing (PBF-EB AM) [64–78]. These findings are important for PBF-EB AM of Ti_6Al_4V, and they may contribute to a better comprehension of the creep behaviour of the alloys.

If AM is to become widely employed in industry, production must keep its costs competitive while still preserving the desirable features necessary to fulfil the purpose for which the product was designed. Therefore, in order to increase our fundamental understanding of AM metal processing, it is necessary to conduct in-depth studies of a variety of chemical and physical phenomena across a wide range of time and length scales (Figure 4.1). When a metal powder is exposed to a laser beam, there is a simultaneous existence of all four states of matter [79]. This is an important element to take into consideration since it results in material interactions that are not evident during regular processing. Rapid thermal cycling can also result in extreme thermal gradients as well as possibly metastable chemical and physical states, both of which can lead to faults in the metallurgical structure of the material. The primary problem with AM of aluminium and the current lack of acceptable alloys have combined to create a barrier to the technology's widespread application. Lasers are widely used in

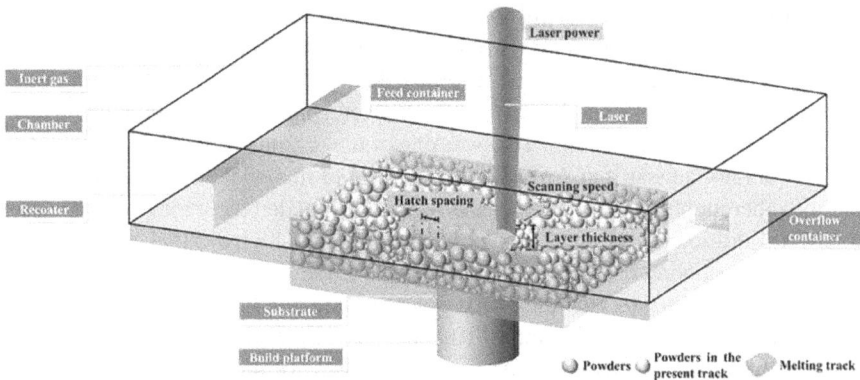

FIGURE 4.1 Typical L-PBF construction chamber and process schematic (black fonts depict the important processing parameters, whereas white fonts show the machine components) [79].

AM of alloys from the powdered form [80]. The objective of this chapter is to provide a synopsis of the many different Al alloys that are presently utilized in (L-PBF)-based AM while stress on the developments that have taken place over the course of the preceding 10 years [81–84]. The main focus is on the mechanical qualities that come from quick solidification as opposed to normal manufacturing and how this affects the microstructural evolution of alloy chemistry. This comparison is made in the light of the fact that the main focus is on the mechanical qualities that come from quick solidification. You can find the information that you are looking for in other recent evaluations; we do not go into great detail on the many AM processing techniques and variables that are required to obtain higher relative densities. Instead, the effect that the process variables have on the microstructure is investigated whenever it is relevant to do so. It has been decided that the only component of the scope that will pertain to the L-PBF approach. In spite of this, it is strongly suggested that the reader looks for additional information in other publications because electron beam melting (EBM) is fully capable of treating aluminium [63].

This chapter summarizes the many types of aluminium alloys currently being used for L-PBF-based AM, with a particular focus on the advancements that have taken place over the past 5 years. The most important aspects of this debate are the influences that traditional manufacturing techniques have on the microstructural evolution of an alloy, as well as the implications that quick solidification has on this process and the subsequent mechanical characteristics.

4.2 ADDITIVE MANUFACTURING (AM) OF ALUMINIUM (AL) ALLOYS

Diverse types of steels, such as austenitic stainless steels and precipitation hardening steels, have generated the most interest in metal AM as the most frequently used engineering material. Closely, the following aluminium and its alloys in this category is the family of materials known as "aluminium alloys". On the other hand, the usage of AM to work with aluminium alloys has been significantly limited up until relatively recently. This resulted in part from the fact that, in contrast to aluminium, alloys are excellently suited for reduced price manufacturing using traditional processes like die casting and machining. This was one of the reasons why this occurred. As a result, it did not appear that AM offered any kind of financial advantage. It appears that the inherent physical features of the Al alloys impeded the widespread adoption of AM. In addition to having a high laser reflectivity and high thermal expansion, the creation of an adherent surface oxide film is one of these properties [85].

It is important to note that the heating and cooling conditions for additively built (AM) components are significantly different from those of conventional manufacturing techniques. As a direct result of this, the order of precipitation phase development and the rate at which it occurs might vary. It has been established that the process of repeatedly heating and cooling a powder feedstock during the printing process can lead to the creation of distinct precipitation phases in the powder. In addition, because AM solidifies so quickly, solutes frequently become trapped within the material. In addition, this makes it easier for the precipitation to become stronger while the patient is having treatment to relieve their stress. It is vital to know the complete AM cycle to regulate the temperature profile and the entire cycle to obtain the ideal precipitation phases and features. AM differs fundamentally from traditional casting and offers substantial benefits due to fewer manufacturing stages and considerably decreased waste [86]. The fundamental L-PBF AM method uses a laser to soften and solidify powder material in coatings in order to construct a three-dimensional structure. This is done in order to generate a 3D microstructure. When coupled with CAD, this method produces excellent results. It is feasible to produce structures and geometries of an unimaginably high level of complexity using ordinary processing methods. Some examples of this include complicated foams, hollow structures, and lattices. As a consequence, this leads to an extremely effective utilization of the available material as well as a good durability and toughness [87]. Metal alloys that solidify rapidly may have unique structural and mechanical properties, which are not achievable with the more conventional method of solidification. Because traditional casting is not allowed for rapid cooling rates throughout the entirety of the cast, this technique can only be used for casting relatively small components. PBF in AM offers a variety of benefits despite the fact that the laser can only heat up very small amounts of material at one time. When combined with brief intervals of laser irradiation, these techniques are capable of producing very rapid changes in temperature, both heating and cooling [87]. This results in significantly variable processing conditions, which in turn produces a metallurgical response that is very dissimilar to that produced by typical casting procedures. Among the most distinguishing features of PBF AM is its ability to rapidly heat and cool the environment. There are three main ways in which the microstructure of an aluminium alloy is altered by fast solidification after the alloy has been created. To begin, quick solidification results in high undercooling, which in turn induces changes in the composition of the material. In more severe situations, this could result in the absence of partitions. Secondly, microstructural refinement is proportional to the pace at which the solidification interface is relocating, and refinement occurs during both bulk and individual phases [87]. Thirdly, even at slow cooling rates, metastable phases can evolve, such as amorphous structures created by some alloys that solidify quickly and quasicrystal line phases that can develop depending on the alloying addition. Both of these can occur even when the cooling rate is quite low. After rapid solidification, improved microstructural characteristics in aluminium alloys are frequently observed. These characteristics include a smaller dendritic arm spacing, fewer amorphous structures, and quasicrystals. Other microstructural improvements include amorphous structures and quasicrystals.

A wide processing window alloy that enables the formation of heterogeneous and hierarchical characteristics when combined with L-PBF has the potential to be utilized in the production of a wide range of components that have varying requirements for their microstructure [88]. The example that follows explains how the microstructural requirements for different product pieces might vary: The vehicle frame requires fine-grained microstructure for improved room-temperature strength, which is achievable at high laser power and rapid scanning speed. On the other hand, the engine features require thermally stable coarse-grained microstructure, which is achievable at low laser power and moderate scanning speed. L-PBF can be utilized for the application-specific production of such Al alloys, as shown in Figure 4.2.

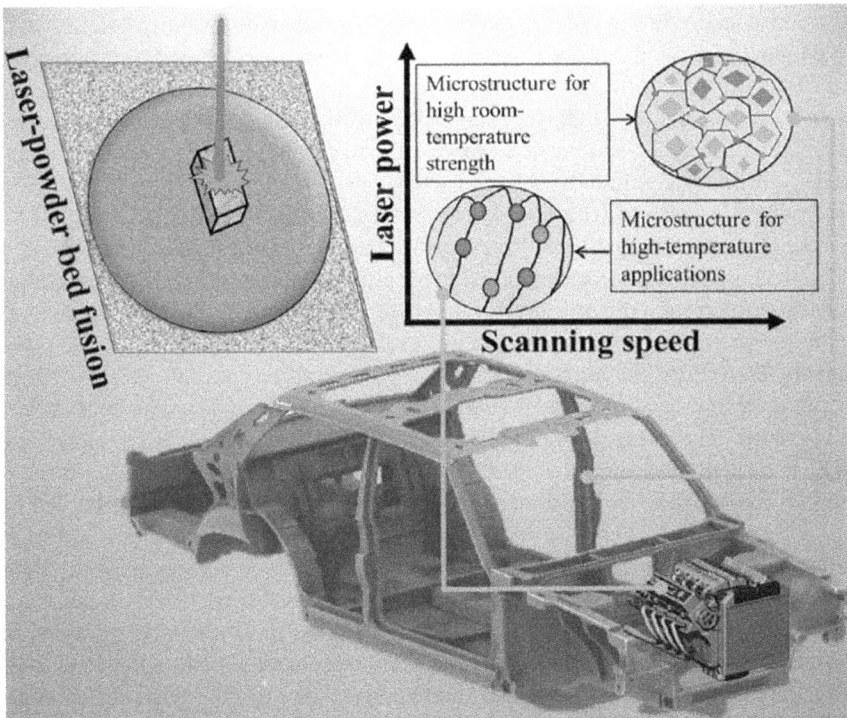

FIGURE 4.2   Achieving application-specific manufacture with L-PBF [88].

4.3 ADDITIVE MANUFACTURING (AM) MICROSTRUCTURE OF AL-SI-BASED METAL MANUFACTURED BY L-PBF

Near-eutectic Al-Si alloys have great heat conductivity and exceptional fluidity. As a result, hypoeutectic Al-Si (9 wt%) alloys make up the common of the Al alloys apply for PBF. The formation of the microstructure at the time of solidification is an extremely important feature in defining the mechanical properties of products that have been processed with L-PBF. In hypoeutectic Al-Si alloys, microstructure plays a significant role in the enhancement of performance of a component. In the tradition of L-PBF, the columnar epitaxial grain is an example of a typical primary-Al grain form. The anisotropic mechanical properties of things manufactured using AM metal are caused by the columnar granules' orientation in a way parallel to the build (in the vertical direction). As a result of the previously hardened layer being partially melted during the material deposition process and spreading over several succeeding building layers, epitaxial columnar grains grow larger. This occurs as a result of the phenomenon known as "epitaxial columnar grain expansion" [89]. This creates a sufficiently strong temperature gradient. The gradient of the melt pool causes latent heat to be released, which in turn prevents new nucleation that takes place just before the solidification process. According to the findings of the EBSD experiment, these columnar grains have a texture similar to fibre. Long and winding columns grain formation occurs when there is a high temperature, and directional heat transfer is caused by sudden changes in temperature in heating and chilling [59]. As a consequence of this, a large number of researchers have examined these effects by employing a wide variety of scanning techniques to enhance the efficiency of L-PBF production, and it is necessary to modify the grain morphology.

In hypereutectic Al-Si alloys, primary-Si can be found in the microstructure of the material. Needles and particles made of eutectic silicon are placed in a matrix made of primary aluminium, with Si particles originating from the primary source generating a material that is very strong and resistant to wear. Because of their low ductility and machining difficulty, primary-Si particles become faceted and "blocky" when they are cast using conventional methods, which severely restrict the applications for which they may be used. As shown in Figure 4.3, these stipulations are binding primary-Si particles that can be resolved by spreading and polishing them over the entirety of the aluminium matrix. Compared to conventional casting, the size of the primary-Si particles produced by AM for alloys containing up to 20 wt% Si is smaller. The internal melt pool of an Al-50Si alloy with a very high percentage of Si is located near the laser medium. It hardened progressively with a decreasing percentage of Si as it moved away from the start. The shallow melt pool has a faster cooling rate than the deeper pool, which results in the primary-Si phase having a smaller size. The primary-Si phase originates as a nucleus from the liquid metal during L-PBF. This microsegregation inside the structure, which is essentially a consequence of the energy input, has a major impact on both the temperature and the size of the melt pool. In addition, the scanning rate as well as specific processing parameters significantly changes the microstructure of the hypereutectic phase, where during L-PBF, quicker cooling rates cause phases to be displaced, which can result in a hypereutectic alloy producing a microstructure that is analogous to that

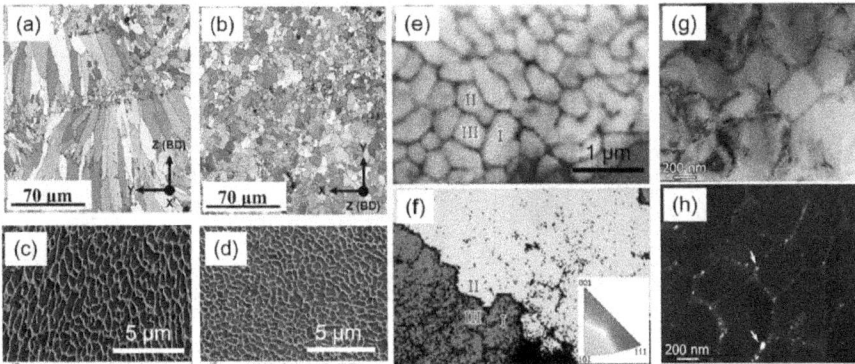

FIGURE 4.3 Microstructural characteristics of the hypereutectic AlSiMg alloy is shown in (a) the EBSD image, while Al grain microstructure is shown in (b) the horizontal orientation. (c) ZY and (d) SEM images demonstrating a fine eutectic microstructure Photos taken at higher magnification that show (e) a band contrast photograph and (f) a directional map, respectively, demonstrate how cell structures inside primary-Al grains are oriented similarly. (g) and (h) are TEM images demonstrating small eutectic Si inside of the cells [92–94].

of a hypoeutectic alloy. Variations in the solidification of the primary phase and the fraction of phases are mostly to blame for this phenomenon [90]. In hypoeutectic compositions, main aluminium is the principal phase because it grows in a directed manner when subjected to a significant temperature gradient and experiences rapid heat transmission. Fine particles of the minor eutectic can be found dispersed throughout the intergranular gaps. On the other hand, hypereutectic alloys have a sizeable amount of eutectic volume because they form when silicon particles are distributed. Even though a columnar microstructure is not going to form in hypereutectic alloys, there is still a possibility that the grains will not be distributed evenly and will segregate in an uneven manner due to the considerable temperature gradient and the associated fluid movement. In motion are the Si particles. The solidification is affected by cooling and the temperature gradient of the fluid flow in the laser processing [91]. Certain factors have an effect on the microstructure of both the hypoeutectic and the hypereutectic phases. Even though their microstructures and compositions are different, aluminium-silicon alloys have the same basic properties in their growth and progress.

4.4 NUMEROUS DEFECTS IDENTIFIED IN THE AM

According to the findings of recent studies, the processing factors have a major effect on the density of AM Al-Si apparatuses, which can be affected by the pore growth of the material. This has a significant and detrimental effect on the overall mechanical design. Processing settings can be increased to boost the construction density in L-PBF. For instance, using powerful lasers, rapid scan rates, and close scan spacing are all examples of how this can be done. However, the influence of the material chemistry on the eventual defect is not nearly as well known.

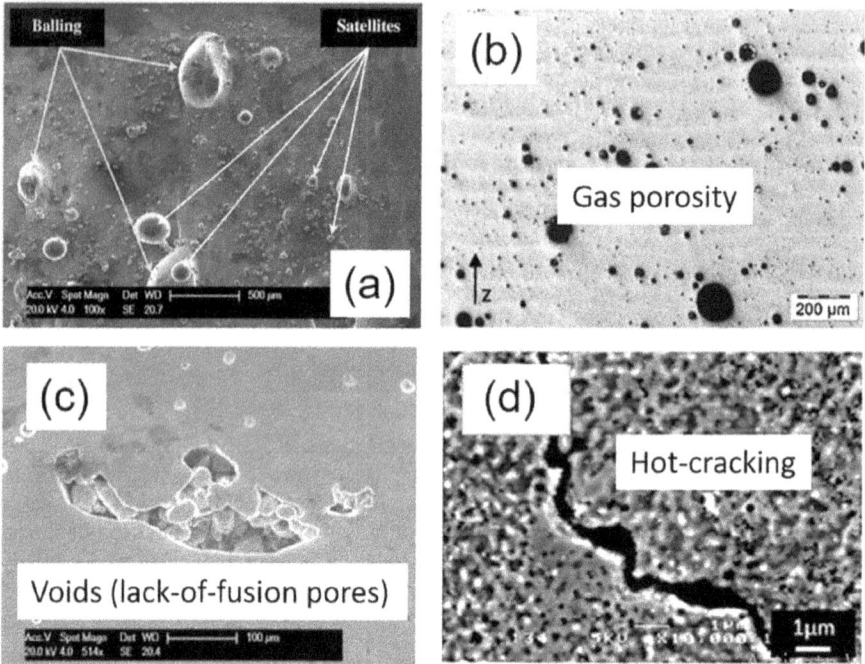

FIGURE 4.4 (a) During the process of L-PBF additive manufacturing of aluminium-Si metal composition is showing balling, (b) gas pores, (c) voids or a lack of fusion pores, and (d) hot cracking are possible to form [95–98].

The formation of AM AlSiMg alloys was discovered to have a number of defects, some of which are depicted in Figure 4.4. The balling phenomenon, depicted in (Figure 4.4a), is frequently observed in L-PBF of alloy metal. This results in a rough scan track and poor image quality. A link is established between the lines. In addition to this, such balling creates an important barrier during the addition of additional force to existing coatings after those layers have already melted. As a consequence, the material becomes porous, lacks consistency, and may even delaminate. Therefore, balling has a considerable negative impact on the geometry and characteristics of the part. Pore flaws with an asymmetrical shape are produced when only a portion of the material is fused, which causes gas to become trapped in specific areas. In contrast, vertically aligned defects are related to a lack of overlap between scan tracks. Because of the presence of moisture in the powder stock, the potential exists for the formation of minute gas pores, also known as "pores". Pore area was found less than 5 m in circumference, seen in Figure 4.4b. This is a really worrisome development, when applying large energy densities. Additionally, the moisture interacts with the released hydrogen, which is capable of being absorbed by the Al, which results in the melting of Al_2O_3. In turn, this results in pores that are rich in hydrogen and get bigger as an L-PBF was found to have generated pores, as stated by Weingarten et al. who constructed $A_1Si_{10}Mg$ alloy consisting of 85% hydrogen. However, research has indicated that pre-drying powders can reduce pore formation; for instance, the build chamber pre-drying method dramatically boosts build density. The keyhole is connected to enormous gas holes. High volumetric energy densities, the manner in which melting occurs, and the utilization of the same processing settings have all been factors in the observations reported in a variety of locales. The core is the innermost volume, while the contour is the outermost edge that defines the shape. Scans are performed within the geometry of a layer. The original location might be found within the contour scan. Places in the powder on one side of the melt show a low heat transfer rate from the liquid.

The pool will cause the temperature to rise. The local energy density was created by changing the scan direction. This led to the acceleration and slowdown of the laser during this time. By optimizing the processing settings for L-PBF of Al-Si alloys, which enables the fabrication of components that are dense even when feedstocks are not pre-treated, this lack of fusion pore defect (Figure 4.4c) may be significantly controlled. Near-eutectic Al-Si alloys are typically not susceptible to solidification cracks (Figure 4.4d), with the exception of situations in which the Si content is present. The hot cracking appeared in L-PBF samples initially as a consequence of shrinkage porosity and then spread as a consequence of lingering stresses that were produced during the construction process.

4.5 MECHANICAL PROPERTIES AFFECTED BY MICROSTRUCTURE

The Si alloys play crucial functions in each of these distinct materials, as is well known. Ca stability of Al-Si alloys ensures the mechanical qualities of the material. The forms of the ordinarily solidified alloys are needle-like or plate-like. During the beginning stages of tensile deformation, the Si phase is responsible for localized shearing, fast initiation, propagation, and accumulation of stresses, as well as plastic deformation to do with breaks and fissures. The nano-sized spherical phase of silicon that is formed at the eutectic areas and in the cells of L-PBF, on the other hand, is resistant to the local shearing stresses. This prevents the formation of cracks and further damage. Strength and flexibility are both increased by propagation. The relevant research indicates that this results in an improvement in the tensile characteristics of AM hypoeutectic Al-Si alloys as compared to material that is cast in the conventional manner [99]. These high tensile characteristics can be attributed to the fact that L-PBF structures have the ability to bend without breaking. Hypereutectic alloys, which are very similar to hypoeutectic alloys, develop strength through the refining of their eutectic and primary-Si phases. This difference in microstructure causes the parts to have anisotropic properties. In spite of the anisotropic properties that can be seen in Al-Si alloys when they are tested in L-PBF, the vast bulk of the research that has been conducted has found that the tensile strengths of Al-Si in both directions are very close to being the same. In addition, it is possible to generate good tensile qualities even when the conditions of the manufacturing are altered. In addition, it was discovered that making changes to the scanning strategy, for instance, by varying the hatch type and contour can drastically change the texture of the material and bring about improvements in its tensile qualities. Different fracture propagation paths are mostly to blame for this phenomenon. It is important to bring attention to the fact that L-PBF samples have a higher hardness. Nevertheless, this effect can be altered by processing elements such as the build and the scan orientation. The vast majority of researchers have brought attention to the properties that L-fatigue PBF possesses. The samples taken from casts are considered to be of lower quality. There is a tensile stress present when it has been revealed that residual tensions, porosity, and unmelted particles can be found in the most likely factors contributing to this. In addition, it has been observed that the heat-affected zone (HAZ) is the location in which fracture occurs the majority of the time at the edge of the melting pool. The size of the HAZ is largely controlled by the melting pool. Adjusting the parameters of the L-PBF processing can be done with relative ease. The difference in temperature between the HAZ and the melt pool, in particular, is steeper. In theory, it is beneficial that gradient decreases the size of the HAZ. According to a study [60] that compared the effects of numerous environments, including Ar, N_2, and He, the samples produced by Ar and N_2 had superior mechanical properties to those produced by He, particularly flexibility. The creation of pore clusters has been proposed to explain this. In addition, the use of the sample's dimensions is a significant factor that plays a significant role in selecting the mechanical properties of the L-PBF-produced sample. Liu et al. [93] studied the effect that changes in the build pace, preheating of the base plate, and post-build heat treatment had on the mechanical properties of $AlSi_{12}$. According to their investigation findings, these aspects impact the material's microstructure and contribute to an increase in the item's resistance to fatigue when incorporated into goods that are created using L-PBF.

Increasing the temperature of the plate made of AlSi10Mg at its base resulted in an increase in grain size and decreased material's degree of hardness. This was useful since it contributed to the end product having a microstructure that contained fewer errors as a result of its production. A part's capacity to adhere to the platform it is being printed on is improved by using a heated build plate, which also helps lower the amount of residual strain and thermal strain that is present. As Si particles are exposed to a variety of conditions, a great deal of attention is devoted to both the evolution of the particles' mechanical characteristics and the morphology of the particles themselves [99]. When it comes to the mechanical properties of Al-Si alloys, the size, form, and distribution of the eutectic Si phase can have a significant amount of influence on those qualities. During the annealing process, the microstructure of the eutectic Si is shown to go through some changes, which are depicted in the following diagram in a schematic form (Figure 4.5). It has been demonstrated that spherical Si particles develop at the grain boundaries of Al when L-PBF is performed because of the high cooling rates that are observed during this process. When the annealing temperature or time is increased, eutectic silicon goes through an expansion process, which results in a decrease in the overall number density. It is believed that the presence of this microstructure is responsible for the high tensile ductility of the AlSiMg alloy, which can reach up to 25%. This great tensile ductility, however, comes at the expense of the material's toughness and strengths [100]. Altering the methods of heat treatment can, on the other hand, allow the grain structure of Al-Si alloys to be modified according to the end user's requirements. In conclusion, the manufacture of hypo- and hypereutectic Al-Si-based alloys is an easy process. This is true for both types of alloys. Throughout the various stages of processing, L-PBF is utilized. The L-PBF processing leads to a rapid solidification of the samples, which considerably

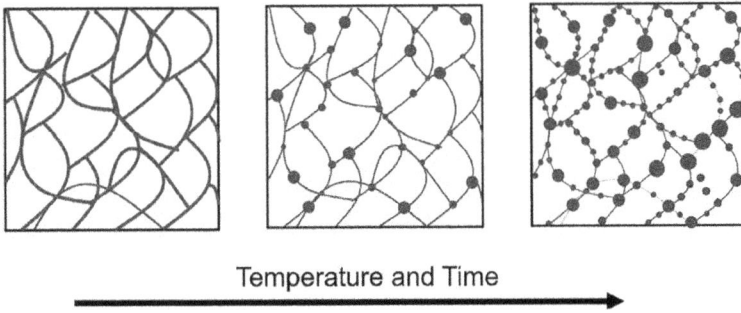

Temperature and Time

FIGURE 4.5 A diagram illustrating the progression of the microstructure of the samples after being treated with PBF and then annealed [102].

enhances the microstructure and raises the level of the material's strength. This is in comparison with more typical casting procedures, which lead to a slower solidification of the samples. In most cases, a material's deteriorating ductility and fatigue qualities may be traced back to the presence of residual porosity, unmelted particles, stresses, and HAZ. However, these same elements can also play a role in the development of new pores in the material. It is possible to increase the ductility of L-PBF-processed samples by applying the necessary heat treatment; however, this will often come at the expense of the samples' strength [61,62,101]. However, it is conceivable to increase the ductility of the samples. It is necessary to keep in mind both the good and negative features of the PBF procedure during the development of components for use in practical applications. This is because the PBF process has both positive and negative characteristics.

4.6 ADDITIVE MANUFACTURING REQUIRES THE REFINING OF GRAIN

During the solidification process of AM, preventing the production of columnar primary-Al grain formations is one of the most critical and difficult obstacles that must be overcome. High-temperature gradients, rapid cooling, and a partial remelting of material that was previously deposited are the distinguishing characteristics of the AM process. These components all play a part in the creation of columnar grains by epitaxial growth, which is the final output of the AM process. Because of the heat stress and the shrinkage that happen during the solidification process, that run between these grains is at a risk of becoming brittle. This is because of the combination of these two factors. Intergranular hot tearing may develop as a consequence of this. In addition, columnar grains lead to the development of mechanical anisotropy, which is a property that is almost always undesirable and should be avoided if possible. It is ideal to end up with a product that has a grain structure that is consistent, fine, and evenly distributed [103,104]. This particular form of grain structure yields structures that are capable of withstanding hot tearing and has isotropic mechanical properties. Since the grain microstructure is built into AM components, it is difficult to modify the overall qualities of the alloys that are produced using this method. In contrast to this, the grain structure of the solidified material in high-strength aluminium alloys that are made using more traditional methods can have its imperfections smoothed out by additional thermo-mechanical processing. As a consequence of this, the strategy that is most useful for AM is to speed up the solidification process while simultaneously controlling the temperature gradient in order to encourage the creation of grains that are correctly equiaxed.

When opposed to earlier powder metallurgical techniques, AM focuses a far more emphasis on the qualities of the powder feedstock that is utilized in the manufacturing process. This is because AM is a more recent manufacturing technique. The usability of the powder in L-PBF is primarily determined by three factors: the size of the powder, the form of the powder, and the dispersion of the powder. These parameters have a direct impact on the powder flow, as well as the mechanical qualities, surface roughness, flaws, packing density, and character of the melt pool. Suppose one desires for the finished build to operate in a dependable and consistent manner. In that case, it is imperative that the quality of the powder be maintained at a uniform level [105]. The characteristics of each powder as well as the ways in which those characteristics influence the product's appearance was studies by conducting a comprehensive literature. The principal production methods for aluminium powders are known as "gas atomization" and "plasma atomization" in environments containing inert gases such as argon, helium, and nitrogen. These atomization processes take place in surroundings. It is likely that gas atomization of aluminium, rather than plasma atomization, is the method that is utilized the most frequently for the purpose of atomizing aluminium. In the long run, plasma-based fragmentation (PBF) will reap the benefits of enhanced sphericity and size uniformity, both of which are discussed in the literature on plasma atomization. In comparison with the raw powder, the spherical particles that emerged as a result exhibited significantly improved flowability and laser absorption [62]. Powders made of aluminium alloys can have their physical characteristics altered by modifying not only the parameters of the atomization method but also the actual atomization processes themselves. The certified vendors, commercially available aluminium alloys, serve as the most accessible powder feedstock for aluminium alloys. This is the case even though alloys such as these are commercially available. One of the few alloys that is capable of being used in the production of aluminium alloys is scalmalloy. The powder forms of only a select few of the most prevalent alloys are available for purchase in commercial quantities. Powder feedstocks can be fairly pricey and do not come in a large variety of forms, which is unfortunate. Because of this condition, utilizing aluminium alloys in AM will be more difficult than it would otherwise be. It is possible to generate a workable alternative by combining a variety of powder feedstocks that are available for purchase in the market in order to produce final goods with the necessary alloy composition. On the other hand, the variability of the composition and the microstructural features that it develops are both undesired outcomes of this process.

Numerous mathematical models have been provided in order to examine how powder packing relates to PSD and to raise packing density. These models have been developed in order to boost packing density. When the packing efficiency of coarse powder matrices is improved, there is typically a subsequent decrease in the amount

Unimodal

Bi-modal

Tri-modal

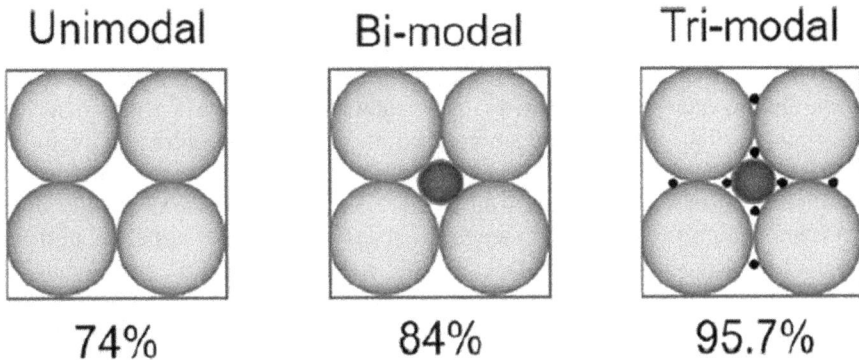

74%   84%   95.7%

FIGURE 4.6   Particle packing density and layout in addition to particle comparing compositions and packing density [106].

of interstitial voids that are present in the matrix. This is the conclusion that was reached from a number of different investigations (Figure 4.6). An increase in the packing efficiency can be achieved by including fine particles, which are able to fill the pores that are present in loose granular networks. It is conceivable to raise the packing density from 74% to 84% by adding minute particles with a size that is the same as the size of its interparticle gaps (Figure 4.6). This would result in a higher packing density overall. The incorporation of a third element has the ability to fill in any gaps that may still be present [106]. As a direct consequence of this, achieving a highly dense manufacturing is not an impossible task.

4.7   CONCLUSIONS

Aluminium is the second furthermost significant material, behind steel, because of the remarkable qualities it possesses. Steel is the most important metal. A stronger and resistance to corrosion are two crucial features to look for in a material. Aluminium is one of the most valuable types of material due to the various advantages it possesses, the low price at which it can be produced, and the simplicity with which it can be fabricated. In contrast to others such as composites and titanium, these materials are considered the most desirable for use in aerospace and automotive applications. Recent efforts on AM with aluminium highlight both the advantages and limitations of this manufacturing technology. In the most recent body of research, near-eutectic AlSiMg alloys have been investigated in great depth, beginning with the feedstock materials and continuing all the way through the functionality of the actual component parts. This examination began at the beginning of the chain of production and moved all the way through to the end. Due to the issue of hot cracking, which is caused by the solidification of the alloy at rapid cooling rates, such as those encountered in AM processing, research on high-strength aluminium alloys is still in its early stages. This is because hot cracking is caused by the rapid cooling rates that are encountered in AM processing. Because of this issue, development has been slowed down on this front. The microstructural refinement that takes place as a result of fast cooling and heat treatment is a significant contributor to the fact that AlSiMg alloys generated by L-PBF have a stronger tensile strength than their counterparts that are produced through cast-based processes. When compared to equilibrium values, the as-printed samples show a higher solute concentration with quick cooling. This leads to a reduction in the amount of time required for treatment utilizing solution heat. Because of its strong heat conductivity and low absorptivity, aluminium is notoriously difficult to melt. As a consequence of this, components with a high vapour pressure are vaporized. It is possible that the loss of these components will cause chemical heterogeneity within the L-PBF-treated sample, in addition to having an effect on the solution's hardness and the precipitate's ability to become more rigid as a result of the loss. Powder qualities such as morphology, packing density, and hydroxides have a significant impact on flowability, produce a variety of defects. Flowability refers to the ability of a powder to be moulded or extruded without rupturing. Flowability is susceptible to being adversely impacted by a variety of powder properties.

4.8 OUTLOOK

In future, additional research will be necessary in order to overcome the challenges that have been presented. The AM has been credited with the discovery of aluminium alloys. In a similar vein, the difficulties are of a scientific kind, as well as science and technology; the Ishikawa report draws attention to a number of topics that are particularly vital, the example that may be found in Figure 4.7. There is a substantial amount of preparatory work that needs to be carried out to create a connection between the field of process metallurgy and the science of the solidification of matter. As a consequence of this, the emphasis of further study ought to be dispersed throughout a number of distinct fields. These alloys go through a process of rapid and thermal cycling when they are subjected to PBF. This process happens frequently

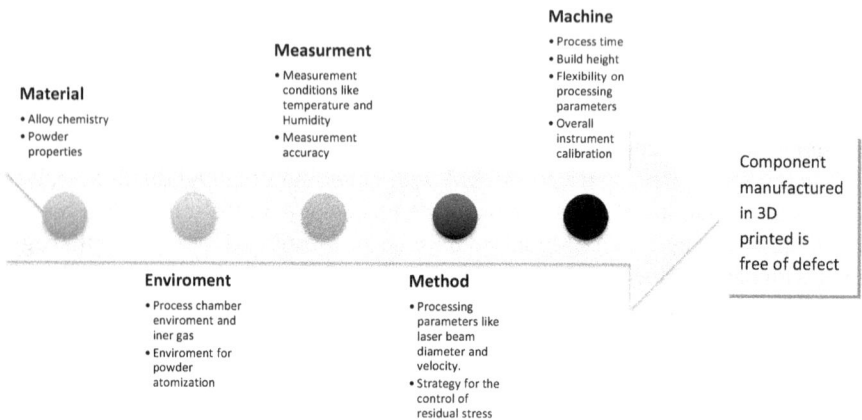

FIGURE 4.7 Highlighting the key technical and scientific difficulties involved in metal additive manufacturing.

and is responsible for a wide variety of defects, like insufficient fusion, and alloying element loss by vaporization. Other defects include persistent pressures and unfavourable microstructural character. The vast bulk of research on high-strength aluminium alloys is concentrated on alloys that are readily available on the commercial market and have been entirely created through a variety of processing techniques. It is of the utmost importance to develop aluminium alloys that are high-strength, high-performance, and cost-effective.

These alloys should make use of the unique attributes that AM possesses in order to offer qualities that are superior to those offered by their traditional analogues. This will make it possible for one to reap the benefits of the rapid solidification that PBF offers. An excellent illustration of this is the rapid rate of cooling, which, when paired with the action of repeated heating, results in an excessive amount of precipitation and the development of dispersoid particles. Another illustration of this is the formation of ice crystals. As a consequence of this procedure, both the purification of grains and the generation of products with superior mechanical qualities are facilitated. In order to enable the use of innovative alloys in order to meet the requirements of the industrial sector in terms of performance and production consistency, it is vital to have an understanding of the links that exist between shape, alloying, processing, and performance.

REFERENCES

1. Prashar, G., Vasudev, H.: A comprehensive review on sustainable cold spray additive manufacturing: state of the art, challenges and future challenges. *J. Clean. Prod.* 310, 127606 (2021). https://doi.org/10.1016/j.jclepro.2021.127606.
2. Prashar, G., Vasudev, H., Bhuddhi, D.: Additive manufacturing: expanding 3D printing horizon in industry 4.0. *Int. J. Interact. Des. Manuf.* (2022). https://doi.org/10.1007/s12008-022-00956-4.
3. Chen, L., Yao, X., Chew, Y., Weng, F., Moon, S.K., Bi, G.: Data-driven adaptive control for laser-based additive manufacturing with automatic controller tuning. *Appl. Sci.* 10, 7967 (2020). https://doi.org/10.3390/app10227967.
4. Mishra, R.S., Palanivel, S.: Building without melting: a short review of friction-based additive manufacturing techniques. *Int. J. Addit. Subtractive Mater. Manuf.* 1, 82 (2017). https://doi.org/10.1504/IJASMM.2017.10003956.
5. Singh, R., Toseef, M., Kumar, J., Singh, J.: Benefits and challenges in additive manufacturing and its applications. In: Kaushal, S., Singh, I., Singh, S., and Gupta, A. (eds.) *Sustainable Advanced Manufacturing and Materials Processing.* pp. 137–157. CRC Press, Boca Raton (2022). https://doi.org/10.1201/9781003269298-8.
6. Sunitha, K., Vasudev, H.: Microsrtructural and mechanical characterization of HVOF-sprayed Ni-based alloy coating. *Int. J. Surf. Eng. Interdiscip. Mater. Sci.* 10, 1–9 (2022). https://doi.org/10.4018/IJSEIMS.298705.
7. Sunitha, K., Vasudev, H.: A short note on the various thermal spray coating processes and effect of post-treatment on Ni-based coatings. *Mater. Today Proc.* 50, 1452–1457 (2022). https://doi.org/10.1016/j.matpr.2021.09.017.
8. Prashar, G., Vasudev, H.: Surface topology analysis of plasma sprayed Inconel625-Al$_2$O$_3$ composite coating. *Mater. Today Proc.* 50, 607–611 (2022). https://doi.org/10.1016/j.matpr.2021.03.090.
9. Singh, J., Vasudev, H., Singh, S.: Performance of different coating materials against high temperature oxidation in boiler tubes – a review. *Mater. Today Proc.* 26, 972–978 (2020). https://doi.org/10.1016/j.matpr.2020.01.156.

10. Singh, J., Kumar, M., Kumar, S., Mohapatra, S.K.: Properties of glass-fiber hybrid composites: a review. *Polym. Plast. Technol. Eng.* 56, 455–469 (2017). https://doi.org/10.10 80/03602559.2016.1233271.
11. Singh, S., Garg, J., Singh, P., Singh, G., Kumar, K.: Effect of hard faced Cr-alloy on abrasive wear of low carbon rotavator blades using design of experiments. *Mater. Today Proc.* 5, 3390–3395 (2018). https://doi.org/10.1016/j.matpr.2017.11.583.
12. Singh, J., Kumar, S., Mohapatra, S.K., Kumar, S.: Shape simulation of solid particles by digital interpretations of scanning electron micrographs using IPA technique. *Mater. Today Proc.* 5, 17786–17791 (2018). https://doi.org/10.1016/j.matpr.2018.06.103.
13. Singh, J., Kumar, S., Mohapatra, S.: Study on role of particle shape in erosion wear of austenitic steel using image processing analysis technique. *Proc. Inst. Mech. Eng. Part J J. Eng. Tribol.* 233, 712–725 (2019). https://doi.org/10.1177/1350650118794698.
14. Singh, J., Singh, J.P., Singh, M., Szala, M.: Computational analysis of solid particle-erosion produced by bottom ash slurry in 90° elbow. *MATEC Web Conf.* 252, 04008 (2019). https://doi.org/0.1051/matecconf/201925204008.
15. Kumar, S., Singh, M., Singh, J., Singh, J.P., Kumar, S.: Rheological characteristics of Uni/Bi-variant particulate iron ore slurry: artificial neural network approach. *J. Min. Sci.* 55, 201–212 (2019). https://doi.org/10.1134/S1062739119025468.
16. Singh, J., Kumar, S., Singh, J.P., Kumar, P., Mohapatra, S.K.: CFD modeling of erosion wear in pipe bend for the flow of bottom ash suspension. *Part. Sci. Technol.* 37, 275–285 (2019). https://doi.org/10.1080/02726351.2017.1364816.
17. Singh, J., Singh, S., Pal Singh, J.: Investigation on wall thickness reduction of hydro-power pipeline underwent to erosion-corrosion process. *Eng. Fail. Anal.* 127, 105504 (2021). https://doi.org/10.1016/j.engfailanal.2021.105504.
18. Ganesh Reddy Majji, B., Vasudev, H., Bansal, A.: A review on the oxidation and wear behavior of the thermally sprayed high-entropy alloys. *Mater. Today Proc.* 50, 1447–1451 (2022). https://doi.org/10.1016/j.matpr.2021.09.016.
19. Singh, J., Singh, J.P.: Numerical analysis on solid particle erosion in elbow of a slurry conveying circuit. *J. Pipeline Syst. Eng. Pract.* 12, 04020070 (2021). https://doi.org/10.1061/(asce)ps.1949-1204.0000518.
20. Singh, J., Singh, S.: Neural network prediction of slurry erosion of heavy-duty pump impeller/casing materials 18Cr-8Ni, 16Cr-10Ni-2Mo, super duplex 24Cr-6Ni-3Mo-N, and grey cast iron. *Wear.* 476, 203741 (2021). https://doi.org/10.1016/j.wear.2021.203741.
21. Singh, J.: Application of thermal spray coatings for protection against erosion, abrasion, and corrosion in hydropower plants and offshore industry. In: Thakur, L. and Vasudev, H. (eds.) *Thermal Spray Coatings.* pp. 243–283. CRC Press, Boca Raton (2021). https://doi.org/10.1201/9781003213185-10.
22. Singh, J.: A review on mechanisms and testing of wear in slurry pumps, pipeline circuits and hydraulic turbines. *J. Tribol.* 143, 1–83 (2021). https://doi.org/10.1115/1.4050977.
23. Singh, J., Singh, S.: A review on machine learning aspect in physics and mechanics of glasses. *Mater. Sci. Eng. B.* 284, 115858 (2022). https://doi.org/10.1016/j.mseb.2022.115858.
24. Kumar, P., Singh, J., Singh, S.: Neural network supported flow characteristics analysis of heavy sour crude oil emulsified by ecofriendly bio-surfactant utilized as a replacement of sweet crude oil. *Chem. Eng. J. Adv.* 11, 100342 (2022). https://doi.org/10.1016/j.ceja.2022.100342.
25. Satyavathi Yedida, V.V., Vasudev, H.: A review on the development of thermal barrier coatings by using thermal spray techniques. *Mater. Today Proc.* 50, 1458–1464 (2022). https://doi.org/10.1016/j.matpr.2021.09.018.
26. Singh, M., Vasudev, H., Kumar, R.: Microstructural characterization of BN thin films using RF magnetron sputtering method. *Mater. Today Proc.* 26, 2277–2282 (2020). https://doi.org/10.1016/j.matpr.2020.02.493.

27. Mehta, A., Vasudev, H., Singh, S.: Recent developments in the designing of deposition of thermal barrier coatings – a review. *Mater. Today Proc.* 26, 1336–1342 (2020). https://doi.org/10.1016/j.matpr.2020.02.271.

28. Sharma, Y., Singh, K.J., Vasudev, H.: Experimental studies on friction stir welding of aluminium alloys. *Mater. Today Proc.* 50, 2387–2391 (2022). https://doi.org/10.1016/j.matpr.2021.10.254.

29. Parkash, J., Saggu, H.S., Vasudev, H.: A short review on the performance of high velocity oxy-fuel coatings in boiler steel applications. *Mater. Today Proc.* 50, 1442–1446 (2022). https://doi.org/10.1016/j.matpr.2021.09.014.

30. Prashar, G., Vasudev, H.: Hot corrosion behavior of super alloys. *Mater. Today Proc.* 26, 1131–1135 (2020). https://doi.org/10.1016/j.matpr.2020.02.226.

31. Singh, J., Kumar, S., Mohapatra, S.K.: Optimization of erosion wear influencing parameters of HVOF sprayed pumping material for coal-water slurry. *Mater. Today Proc.* 5, 23789–23795 (2018). https://doi.org/10.1016/j.matpr.2018.10.170.

32. Singh, J., Kumar, S., Singh, G.: Taguchi's approach for optimization of tribo-resistance parameters Forss304. *Mater. Today Proc.* 5, 5031–5038 (2018). https://doi.org/10.1016/j.matpr.2017.12.081.

33. Singh, J.: Tribo-performance analysis of HVOF sprayed 86WC-10Co4Cr & Ni-Cr$_2$O$_3$ on AISI 316L steel using DOE-ANN methodology. *Ind. Lubr. Tribol.* 73, 727–735 (2021). https://doi.org/10.1108/ILT-04-2020-0147.

34. Singh, J., Singh, J.P.: Performance analysis of erosion resistant Mo2C reinforced WC-CoCr coating for pump impeller with Taguchi's method. *Ind. Lubr. Tribol.* 74, 431–441 (2022). https://doi.org/10.1108/ILT-05-2020-0155.

35. Singh, J., Singh, S., Verma, A.: Artificial intelligence in use of ZrO$_2$ material in biomedical science. *J. Electrochem. Sci. Eng.* (2022). https://doi.org/10.5599/jese.1498.

36. Singh, J., Singh, S., Gill, R.: Applications of biopolymer coatings in biomedical engineering. *J. Electrochem. Sci. Eng.* (2022). https://doi.org/10.5599/jese.1460.

37. Singh, J., Singh, S.: Neural network supported study on erosive wear performance analysis of Y$_2$O$_3$/WC-10Co4Cr HVOF coating. *J. King Saud Univ. - Eng. Sci.* (2022). https://doi.org/10.1016/j.jksues.2021.12.005.

38. Singh, M., Vasudev, H., Kumar, R.: Corrosion and tribological behaviour of BN thin films deposited using magnetron sputtering. *Int. J. Surf. Eng. Interdiscip. Mater. Sci.* 9, 24–39 (2021). https://doi.org/10.4018/IJSEIMS.2021070102.

39. Prashar, G., Vasudev, H.: High temperature erosion behavior of plasma sprayed Al$_2$O$_3$ coating on AISI-304 stainless steel. *World J. Eng.* 18, 760–766 (2021). https://doi.org/10.1108/WJE-10-2020-0476.

40. Singh, J.: Analysis on suitability of HVOF sprayed Ni-20Al, Ni-20Cr and Al-20Ti coatings in coal-ash slurry conditions using artificial neural network model. *Ind. Lubr. Tribol.* 71, 972–982 (2019). https://doi.org/10.1108/ILT-12-2018-0460.

41. Singh, J., Kumar, S., Mohapatra, S.K.: Tribological performance of Yttrium (III) and Zirconium (IV) ceramics reinforced WC–10Co4Cr cermet powder HVOF thermally sprayed on X2CrNiMo-17-12-2 steel. *Ceram. Int.* 45, 23126–23142 (2019). https://doi.org/10.1016/j.ceramint.2019.08.007.

42. Singh, J.: *Investigation on Slurry Erosion of Different Pumping Materials and Coatings.* Thapar Institute of Engineering and Technology, Patiala, India (2019).

43. Singh, J., Kumar, S., Mohapatra, S.K.: An erosion and corrosion study on thermally sprayed WC-Co-Cr powder synergized with Mo$_2$C/Y$_2$O$_3$/ZrO$_2$ feedstock powders. *Wear.* 438–439, 102751 (2019). https://doi.org/10.1016/j.wear.2019.01.082.

44. Singh, J., Kumar, S., Mohapatra, S.K.: Erosion wear performance of Ni-Cr-O and NiCrBSiFe-WC(Co) composite coatings deposited by HVOF technique. *Ind. Lubr. Tribol.* 71, 610–619 (2019). https://doi.org/10.1108/ILT-04-2018-0149.

45. Singh, J., Kumar, S., Mohapatra, S.K.: Erosion tribo-performance of HVOF deposited Stellite-6 and Colmonoy-88 micron layers on SS-316L. *Tribol. Int.* 147, 105262 (2020). https://doi.org/10.1016/j.triboint.2018.06.004.

46. Singh, J.: Slurry erosion performance analysis and characterization of high-velocity oxy-fuel sprayed Ni and Co hardsurfacing alloy coatings. *J. King Saud Univ. - Eng. Sci.* (2021). https://doi.org/10.1016/j.jksues.2021.06.009.

47. Singh, J.: Wear performance analysis and characterization of HVOF deposited Ni–20Cr$_2$O$_3$, Ni–30Al$_2$O$_3$, and Al$_2$O$_3$–13TiO$_2$ coatings. *Appl. Surf. Sci. Adv.* 6, 100161 (2021). https://doi.org/10.1016/j.apsadv.2021.100161.

48. Singh, G., Vasudev, H., Bansal, A., Vardhan, S., Sharma, S.: Microwave cladding of Inconel-625 on mild steel substrate for corrosion protection. *Mater. Res. Express.* 7, 026512 (2020). https://doi.org/10.1088/2053-1591/ab6fa3.

49. Singh, G., Vasudev, H., Bansal, A., Vardhan, S.: Influence of heat treatment on the microstructure and corrosion properties of the Inconel-625 clad deposited by microwave heating. *Surf. Topogr. Metrol. Prop.* 9, 025019 (2021). https://doi.org/10.1088/2051-672X/abfc61.

50. Bansal, A., Vasudev, H., Sharma, A.K., Kumar, P.: Investigation on the effect of post weld heat treatment on microwave joining of the Alloy-718 weldment. *Mater. Res. Express.* 6, 086554 (2019). https://doi.org/10.1088/2053-1591/ab1d9a.

51. Vasudev, H., Singh, P., Thakur, L., Bansal, A.: Mechanical and microstructural characterization of microwave post processed Alloy-718 coating. *Mater. Res. Express.* 6, 1265f5 (2020). https://doi.org/10.1088/2053-1591/ab66fb.

52. Prashar, G., Vasudev, H., Thakur, L.: Influence of heat treatment on surface properties of HVOF deposited WC and Ni-based powder coatings: a review. *Surf. Topogr. Metrol. Prop.* 9, 043002 (2021). https://doi.org/10.1088/2051-672X/ac3a52.

53. Singh, P., Vasudev, H., Bansal, A.: Effect of post-heat treatment on the microstructural, mechanical, and bioactivity behavior of the microwave-assisted alumina-reinforced hydroxyapatite cladding. *Proc. Inst. Mech. Eng. Part E J. Process Mech. Eng.* 095440892211161 (2022). https://doi.org/10.1177/09544089221116168.

54. Vasudev, H., Thakur, L., Singh, H., Bansal, A.: A study on processing and hot corrosion behaviour of HVOF sprayed Inconel718-nano Al$_2$O$_3$ coatings. *Mater. Today Commun.* 25, 101626 (2020). https://doi.org/10.1016/j.mtcomm.2020.101626.

55. Prashar, G., Vasudev, H., Thakur, L.: High-temperature oxidation and erosion resistance of ni-based thermally-sprayed coatings used in power generation machinery: a review. *Surf. Rev. Lett.* 29, 2230003 (2022). https://doi.org/10.1142/S0218625X22300039.

56. Singh, P., Bansal, A., Vasudev, H., Singh, P.: In situ surface modification of stainless steel with hydroxyapatite using microwave heating. *Surf. Topogr. Metrol. Prop.* 9, 035053 (2021). https://doi.org/10.1088/2051-672X/ac28a9.

57. Prashar, G., Vasudev, H.: A review on the influence of process parameters and heat treatment on the corrosion performance of Ni-based thermal spray coatings. *Surf. Rev. Lett.* 29, 1–18 (2022). https://doi.org/10.1142/S0218625X22300015.

58. Vasudev, H., Singh, G., Bansal, A., Vardhan, S., Thakur, L.: Microwave heating and its applications in surface engineering: a review. *Mater. Res. Express.* 6, 102001 (2019). https://doi.org/10.1088/2053-1591/ab3674.

59. Zhang, Y., Yang, S., Zhao, Y.F.: Manufacturability analysis of metal laser-based powder bed fusion additive manufacturing—a survey. *Int. J. Adv. Manuf. Technol.* 110, 57–78 (2020). https://doi.org/10.1007/s00170-020-05825-6.

60. Mollamahmutoglu, M., Yilmaz, O.: Volumetric heat source model for laser-based powder bed fusion process in additive manufacturing. *Therm. Sci. Eng. Prog.* 25, 101021 (2021). https://doi.org/10.1016/j.tsep.2021.101021.

61. Flores Ituarte, I., Wiikinkoski, O., Jansson, A.: Additive manufacturing of polypropylene: a screening design of experiment using laser-based powder bed fusion. *Polymers.* 10, 1293 (2018). https://doi.org/10.3390/polym10121293.

62. Wei, C., Chueh, Y.-H., Zhang, X., Huang, Y., Chen, Q., Li, L.: Easy-to-remove composite support material and procedure in additive manufacturing of metallic components using multiple material laser-based powder bed fusion. *J. Manuf. Sci. Eng.* 141, (2019). https://doi.org/10.1115/1.4043536.

63. Jadhav, S.D., Goossens, L.R., Kinds, Y., Van Hooreweder, B., Vanmeensel, K.: Laser-based powder bed fusion additive manufacturing of pure copper. *Addit. Manuf.* 42, 101990 (2021). https://doi.org/10.1016/j.addma.2021.101990.

64. Usca, Ü.A., Uzun, M., Şap, S., Giasin, K., Pimenov, D.Y., Prakash, C.: Determination of machinability metrics of AISI 5140 steel for gear manufacturing using different cooling/lubrication conditions. *J. Mater. Res. Technol.* (2022). https://doi.org/10.1016/j.jmrt.2022.09.067.

65. Nguyen, D.N., Le Chau, N., Dao, T.P., Prakash, C., Singh, S.: Experimental study on polishing process of cylindrical roller bearings. *Meas. Control.* 52, 1272–1281 (2019). https://doi.org/10.1177/0020294019864395.

66. Prakash, C., Kansal, H.K., Pabla, B.S., Puri, S.: Processing and characterization of novel biomimetic nanoporous bioceramic surface on β-Ti implant by powder mixed electric discharge machining. *J. Mater. Eng. Perform.* 24, 3622–3633 (2015). https://doi.org/10.1007/s11665-015-1619-6.

67. Prakash, C., Kansal, H.K., Pabla, B.S., Puri, S.: Powder Mixed electric discharge machining: an innovative surface modification technique to enhance fatigue performance and bioactivity of β-Ti implant for orthopedics application. *J. Comput. Inf. Sci. Eng.* 16, (2016). https://doi.org/10.1115/1.4033901.

68. Prakash, C., Uddin, M.S.: Surface modification of β-phase Ti implant by hydroaxyapatite mixed electric discharge machining to enhance the corrosion resistance and in-vitro bioactivity. *Surf. Coatings Technol.* 326, 134–145 (2017). https://doi.org/10.1016/j.surfcoat.2017.07.040.

69. Vasudev, H., Thakur, L., Singh, H., Bansal, A.: Effect of addition of Al_2O_3 on the high-temperature solid particle erosion behaviour of HVOF sprayed Inconel-718 coatings. *Mater. Today Commun.* 30, 103017 (2022). https://doi.org/10.1016/j.mtcomm.2021.103017.

70. Vasudev, H., Thakur, L., Singh, H., Bansal, A.: An investigation on oxidation behaviour of high velocity oxy-fuel sprayed Inconel718-Al_2O_3 composite coatings. *Surf. Coatings Technol.* 393, 125770 (2020). https://doi.org/10.1016/j.surfcoat.2020.125770.

71. Prashar, G., Vasudev, H.: Structure-property correlation and high-temperature erosion performance of Inconel625-Al_2O_3 plasma-sprayed bimodal composite coatings. *Surf. Coatings Technol.* 439, 128450 (2022). https://doi.org/10.1016/j.surfcoat.2022.128450.

72. Prakash, C., Kansal, H.K., Pabla, B., Puri, S., Aggarwal, A.: Electric discharge machining – A potential choice for surface modification of metallic implants for orthopedic applications: a review. *Proc. Inst. Mech. Eng. Part B J. Eng. Manuf.* 230, 331–353 (2016). https://doi.org/10.1177/0954405415579113.

73. Tiwari, A.K., Kumar, A., Kumar, N., Prakash, C.: Investigation on micro-residual stress distribution near hole using nanoindentation: effect of drilling speed. *Meas. Control.* 52, 1252–1263 (2019). https://doi.org/10.1177/0020294019858107.

74. Prakash, C., Singh, S., Pruncu, C., Mishra, V., Królczyk, G., Pimenov, D., Pramanik, A.: Surface modification of Ti-6Al-4V alloy by electrical discharge coating process using partially sintered Ti-Nb electrode. *Materials.* 12, 1006 (2019). https://doi.org/10.3390/ma12071006.

75. Prakash, C., Kansal, H.K., Pabla, B.S., Puri, S.: Experimental investigations in powder mixed electric discharge machining of Ti–35Nb–7Ta–5Zrβ-titanium alloy. *Mater. Manuf. Process.* 32, 274–285 (2017). https://doi.org/10.1080/10426914.2016.1198018.

76. Pradhan, S., Singh, S., Prakash, C., Królczyk, G., Pramanik, A., Pruncu, C.I.: Investigation of machining characteristics of hard-to-machine Ti-6Al-4V-ELI alloy for biomedical applications. *J. Mater. Res. Technol.* 8, 4849–4862 (2019). https://doi.org/10.1016/j.jmrt.2019.08.033.

77. Pramanik, A., Basak, A.K., Littlefair, G., Debnath, S., Prakash, C., Singh, M.A., Marla, D., Singh, R.K.: Methods and variables in Electrical discharge machining of titanium alloy – A review. *Heliyon.* 6, e05554 (2020). https://doi.org/10.1016/j.heliyon.2020.e05554.

78. Prakash, C., Kansal, H.K., Pabla, B.S., Puri, S.: Multi-objective optimization of powder mixed electric discharge machining parameters for fabrication of biocompatible layer on β-Ti alloy using NSGA-II coupled with Taguchi based response surface methodology. *J. Mech. Sci. Technol.* 30, 4195–4204 (2016). https://doi.org/10.1007/s12206-016-0831-0.

79. Yu, W.H., Sing, S.L., Chua, C.K., Kuo, C.N., Tian, X.L.: Particle-reinforced metal matrix nanocomposites fabricated by selective laser melting: a state of the art review. *Prog. Mater. Sci.* 104, 330–379 (2019). https://doi.org/10.1016/j.pmatsci.2019.04.006.

80. Kumar, D., Yadav, R., Singh, J.: Evolution and adoption of microwave claddings in modern engineering applications. In: *Advances in Microwave Processing for Engineering Materials.* pp. 134–153. CRC Press, Boca Raton (2022). https://doi.org/10.1201/9781003248743-8.

81. Behera, D., Chizari, S., Shaw, L.A., Porter, M., Hensleigh, R., Xu, Z., Roy, N.K., Connolly, L.G., Zheng, X. (Rayne), Saha, S., Hopkins, J.B., Cullinan, M.A.: Current challenges and potential directions towards precision microscale additive manufacturing – Part II: laser-based curing, heating, and trapping processes. *Precis. Eng.* 68, 301–318 (2021). https://doi.org/10.1016/j.precisioneng.2020.12.012.

82. Molotnikov, A., Kingsbury, A., Brandt, M.: Current state and future trends in laser powder bed fusion technology. In: *Fundamentals of Laser Powder Bed Fusion of Metals.* pp. 621–634. Elsevier (2021). https://doi.org/10.1016/B978-0-12-824090-8.00011-1.

83. Kruth, J.-P., Levy, G., Klocke, F., Childs, T.H.C.: Consolidation phenomena in laser and powder-bed based layered manufacturing. *CIRP Ann.* 56, 730–759 (2007). https://doi.org/10.1016/j.cirp.2007.10.004.

84. Ahsan, F., Ladani, L.: Temperature profile, bead geometry, and elemental evaporation in laser powder bed fusion additive manufacturing process. *JOM.* 72, 429–439 (2020). https://doi.org/10.1007/s11837-019-03872-3.

85. Thampy, V., Fong, A.Y., Calta, N.P., Wang, J., Martin, A.A., Depond, P.J., Kiss, A.M., Guss, G., Xing, Q., Ott, R.T., van Buuren, A., Toney, M.F., Weker, J.N., Kramer, M.J., Matthews, M.J., Tassone, C.J., Stone, K.H.: Subsurface cooling rates and microstructural response during laser based metal additive manufacturing. *Sci. Rep.* 10, 1981 (2020). https://doi.org/10.1038/s41598-020-58598-z.

86. Storck, S.M., McCue, I.D., Montalbano, T.J., Nimer, S.M., Peitsch, C.M.: Metal matrix composites synthesized with laser-based additive manufacturing. *Johns Hopkins APL Tech. Dig.* 35, 16–18 (2021).

87. Kaligar, A.B., Kumar, H.A., Ali, A., Abuzaid, W., Egilmez, M., Alkhader, M., Abed, F., Alnaser, A.S.: Femtosecond laser-based additive manufacturing: current status and perspectives. *Quantum Beam Sci.* 6, 5 (2022). https://doi.org/10.3390/qubs6010005.

88. Thapliyal, S., Shukla, S., Zhou, L., Hyer, H., Agrawal, P., Agrawal, P., Komarasamy, M., Sohn, Y., Mishra, R.S.: Design of heterogeneous structured Al alloys with wide processing window for laser-powder bed fusion additive manufacturing. *Addit. Manuf.* 42, 102002 (2021). https://doi.org/10.1016/j.addma.2021.102002.

89. Aboutaleb, A.M., Bian, L., Shamsaei, N., Thompson, S.M.: Systematic optimization of Laser-based additive manufacturing for multiple mechanical properties. In: *2016 IEEE International Conference on Automation Science and Engineering (CASE).* pp. 780–785. IEEE (2016). https://doi.org/10.1109/COASE.2016.7743481.

90. Sahasrabudhe, H., Bandyopadhyay, A.: Laser-based additive manufacturing of zirconium. *Appl. Sci.* 8, 393 (2018). https://doi.org/10.3390/app8030393.

91. Stavropoulos, P., Papacharalampopoulos, A., Stavridis, J., Sampatakakis, K.: A three-stage quality diagnosis platform for laser-based manufacturing processes. *Int. J. Adv. Manuf. Technol.* 110, 2991–3003 (2020). https://doi.org/10.1007/s00170-020-05981-9.

92. Wu, J., Wang, X.Q., Wang, W., Attallah, M.M., Loretto, M.H.: Microstructure and strength of selectively laser melted AlSi10Mg. *Acta Mater.* 117, 311–320 (2016). https://doi.org/10.1016/j.actamat.2016.07.012.

93. Liu, X., Zhao, C., Zhou, X., Shen, Z., Liu, W.: Microstructure of selective laser melted AlSi10Mg alloy. *Mater. Des.* 168, 107677 (2019). https://doi.org/10.1016/j.matdes.2019.107677.

94. Ch, S.R., Raja, A., Nadig, P., Jayaganthan, R., Vasa, N.J.: Influence of working environment and built orientation on the tensile properties of selective laser melted AlSi10Mg alloy. *Mater. Sci. Eng. A.* 750, 141–151 (2019). https://doi.org/10.1016/j.msea.2019.01.103.

95. Aboulkhair, N.T., Everitt, N.M., Ashcroft, I., Tuck, C.: Reducing porosity in AlSi10Mg parts processed by selective laser melting. *Addit. Manuf.* 1–4, 77–86 (2014). https://doi.org/10.1016/j.addma.2014.08.001.

96. Aboulkhair, N.T., Maskery, I., Tuck, C., Ashcroft, I., Everitt, N.M.: On the formation of AlSi10Mg single tracks and layers in selective laser melting: microstructure and nano-mechanical properties. *J. Mater. Process. Technol.* 230, 88–98 (2016). https://doi.org/10.1016/j.jmatprotec.2015.11.016.

97. Kimura, T., Nakamoto, T., Mizuno, M., Araki, H.: Effect of silicon content on densification, mechanical and thermal properties of Al-xSi binary alloys fabricated using selective laser melting. *Mater. Sci. Eng. A.* 682, 593–602 (2017). https://doi.org/10.1016/j.msea.2016.11.059.

98. Weingarten, C., Buchbinder, D., Pirch, N., Meiners, W., Wissenbach, K., Poprawe, R.: Formation and reduction of hydrogen porosity during selective laser melting of AlSi10Mg. *J. Mater. Process. Technol.* 221, 112–120 (2015). https://doi.org/10.1016/j.jmatprotec.2015.02.013.

99. Stathatos, E., Vosniakos, G.-C.: Efficient temperature regulation through power optimization for arbitrary paths in laser based additive manufacturing. *CIRP J. Manuf. Sci. Technol.* 33, 133–142 (2021). https://doi.org/10.1016/j.cirpj.2021.03.008.

100. Jang, T.-S., Kim, D., Han, G., Yoon, C.-B., Jung, H.-D.: Powder based additive manufacturing for biomedical application of titanium and its alloys: a review. *Biomed. Eng. Lett.* 10, 505–516 (2020). https://doi.org/10.1007/s13534-020-00177-2.

101. Wei, H.L., Mukherjee, T., DebRoy, T.: Grain growth modeling for additive manufacturing of nickel based superalloys. In: *Proceedings of the 6th International Conference on Recrystallization and Grain Growth (ReX&GG 2016).* pp. 265–269. Springer International Publishing, Cham (2016). https://doi.org/10.1007/978-3-319-48770-0_39.

102. Prashanth, K.G., Scudino, S., Klauss, H.J., Surreddi, K.B., Löber, L., Wang, Z., Chaubey, A.K., Kühn, U., Eckert, J.: Microstructure and mechanical properties of Al–12Si produced by selective laser melting: effect of heat treatment. *Mater. Sci. Eng. A.* 590, 153–160 (2014). https://doi.org/10.1016/j.msea.2013.10.023.

103. Harke, K.J., Calta, N., Tringe, J., Stobbe, D.: Laser-based ultrasound interrogation of surface and sub-surface features in advanced manufacturing materials. *Sci. Rep.* 12, 3309 (2022). https://doi.org/10.1038/s41598-022-07261-w.

104. Jones, J.B.: Repurposing mainstream CNC machine tools for laser-based additive manufacturing. Presented at the April 6 (2016). https://doi.org/10.1117/12.2217901.

105. Gao, H., Yang, J., Jin, X., Zhang, D., Zhang, S., Zhang, F., Chen, H.: Static compressive behavior and failure mechanism of tantalum scaffolds with optimized periodic lattice fabricated by laser-based additive manufacturing. *3D Print. Addit. Manuf.* (2022). https://doi.org/10.1089/3dp.2021.0253.

106. Zhu, H.H., Fuh, J.Y.H., Lu, L.: The influence of powder apparent density on the density in direct laser-sintered metallic parts. *Int. J. Mach. Tools Manuf.* 47, 294–298 (2007). https://doi.org/10.1016/j.ijmachtools.2006.03.019.

5 Laser Surface Engineering
State of the Art, Applications, and Challenges

Gaurav Prashar
Rayat Bahra Institute of Engineering and Nano Technology

Hitesh Vasudev
Lovely Professional University

CONTENTS

5.1 INTRODUCTION

Materials, information, and energy serve as the three cornerstones of contemporary societal progress. Materials serve as the foundation for energy and information in these three pillars. The advancement of human society is intimately tied to the development of materials and the technology used in their manufacture [1–6]. Recently, laser surface engineering (LSE) technology has gained a lot of popularity for its novel forming method. LSE is better suitable for treating the surface of metallic substrates, light alloys regarding process cost, flexibility, and convenience of use, particularly for the creation of a wide variety of exotic compositional and/or microstructural alterations [7]. The specific region of the substrate/component microstructure and/or composition may be altered by heating, melting, deposition, alloying, or cladding via a laser beam, which is a clean heat source [8–16]. The general classification of LSE is shown in Figure 5.1. The three main categories of operations are heating alone (without melting), melting alone (without vaporizing), and vaporizing. Figure 5.2 shows

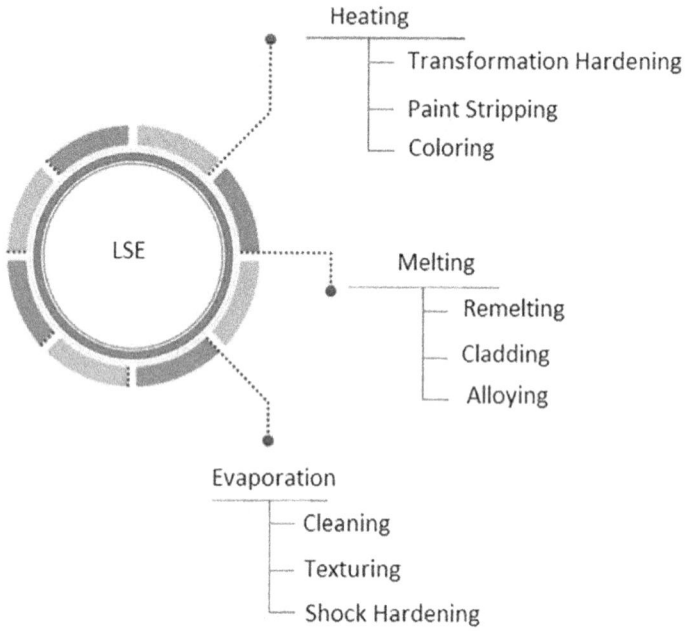

FIGURE 5.1   General classification of LSE.

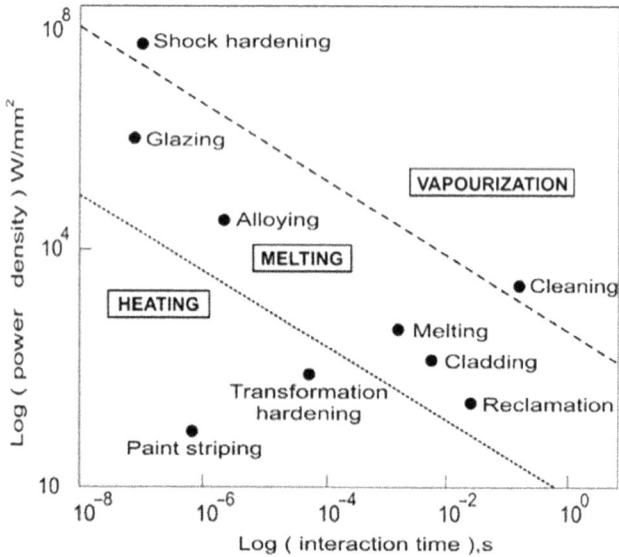

FIGURE 5.2   Processing regime of various LSE approaches [17].

the regime for a single LSE technique as a function of the applied power density and contact time. It should be highlighted that surface heating and low power density are all that are needed for transformation hardening. Surface melting, cladding, glazing, and reclamation all entail melting and demand a medium-high power density. On the other hand, a very high power density is required for cleaning and shock hardening as both call for the elimination of materials as vapour [17]. Therefore, it is necessary to carefully regulate and optimize a number of crucial process parameters in order to obtain the ideal surface microstructure and composition. The type of laser being utilized (laser wavelength), power density, size of beam, geometry of workpiece, working distance, feed rate, and angle of deposition are the primary process variables that actively regulate the microstructure and composition of the surface. Applications for LSE have been developed using Nd:YAG, CO_2, and excimer lasers.

5.2 SURFACE ENGINEERING FOR WEAR AND CORROSION RESISTANCE

It is common knowledge that the corrosion causes a variety of materials to deteriorate through chemical interactions with their surroundings [10–16,32,18–55]. Most corrosion is electrochemical in nature, where electrons are transported from the metal surface to a liquid electrolyte solution. The metal surface is oxidized as a result of these reactions, and corrosion may result in system failure. In this scenario, it is anticipated that corrosion-related costs will account for 3%–5% of the GNP (gross national product) of developed nations. Corrosion-related structural component damage is projected to cost more than US$1.8 trillion annually [56]. Surface engineering seeks to enhance surface-dependent engineering features by altering the microstructural features and/or compositions of a specific part [8,9,57–74]. For the alteration of the microstructure and/or surface composition of any part, equilibrium or near-equilibrium treatments (such as heat treatment, surface coatings, and painting) are typically used [75–85]. Traditional surface engineering techniques use a variety of methodologies and cover a wide application area. The below-mentioned qualities or properties of components with surface engineering are desired:

- increased corrosion resistance via sacrificial protection,
- enhanced oxidation or sulphidation resistance,
- resistance to wear,
- better mechanical properties,
- improved thermal insulation.

There are many surface engineering approaches available to attain these properties. The best choice will depend on a number of factors, including component size and accessibility, corrosive environment, predicted temperatures, component deformation, coating thickness (Figure 5.3), and costs.

Although choosing a surface engineering process and material may appear challenging, it is typically simple. The availability of options frequently places restrictions on the choice (for example, laser melting and/or alloying can only be obtained

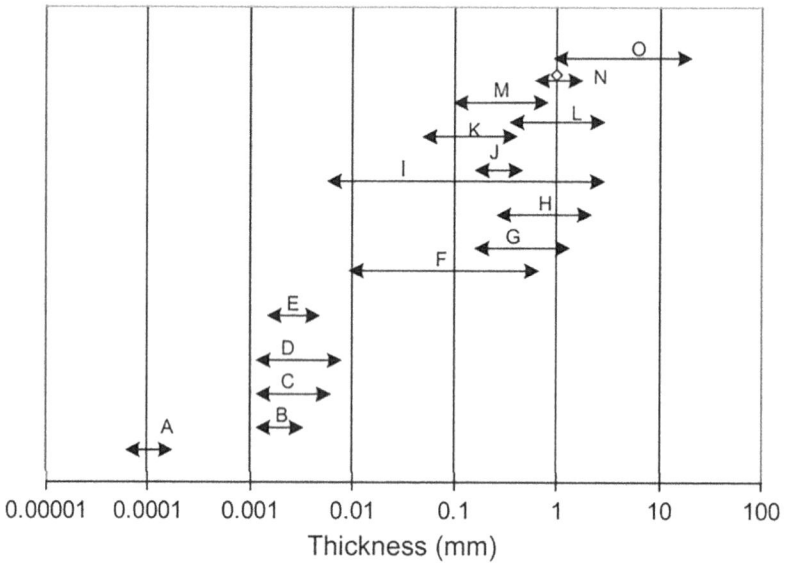

FIGURE 5.3   Approximate thickness of various surface engineering techniques: (where A, ion implantation; B, PVD; C, CVD; D, electrolytic plating; E, electroless plating; F, hot dipping (galvanizing/aluminizing); G, laser surface alloying; H, transformation hardening; I, mechanical working; J, nitriding; K, carbonitriding; L, carburizing; M, thermal spraying; N, friction surfacing; O, weld overlays) [86].

through a particular arrangement with laser job shops). Although there is frequently a precedent, it can be helpful to use a checklist when thinking about a new challenge. The judgements that must be made in order address various important issues. Firstly, it is important to have a firm understanding of service circumstances based on past experience or plant data. For instance, various types of materials can be treated using laser surface modification without mechanical touch and with a low risk of contamination. Additionally, this photonic technology is a viable tool in the context of modern manufacturing due to its simplicity of automation and scalability. The photonics scene is continually being updated as more effective novel solutions attain high degrees of technological readiness as a result of all these distinct traits. The next section discusses some studies on wear and corrosion resistance properties of different materials processed by LSE technology.

5.3   LSE: CORROSION AND WEAR STUDIES

In the past, LSE was successfully applied on many materials to enhance their wear and corrosion resistance properties. Microstructural alteration via LSE was accomplished by melting the surface and quickly cooling it (e.g., laser surface melting). The LSE experimental setup used by Park et al. [87] to examine the corrosion and wear resistance of **mould steel** before and after LSE is shown in Figure 5.4. For the LSE, a 4 kWh diode laser with a continual wave mode and a wavelength of 970 nm was

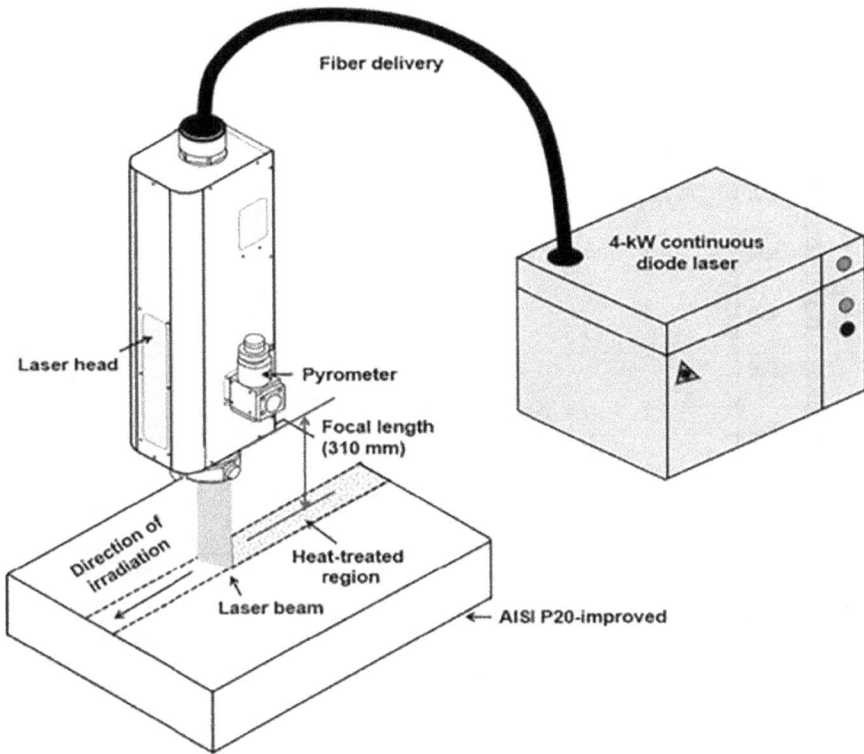

FIGURE 5.4   Experimental setup employed for the LSE of P20-improved steel [87].

used. The authors concluded that trend in improved wear resistance followed the rise in surface hardness brought on by LSE at various laser energy densities as shown in Figure 5.5. Furthermore, the increase in the microstructural uniformity offered by the LSE was what caused the corrosion resistance to improve. Table 5.1 summarizes the wear and corrosion properties of laser surface-treated improved steels and base material.

Mg-based alloys continue to have relatively inferior surface qualities, like wear and corrosion protection, despite great progress being made in modifying the bulk properties of Mg, like formability and creep resistance. The production of undesired brittle intermetallic phases and elemental segregation limit the improvements that alloying can make to wear and corrosion characteristics. Surface engineering techniques are frequently used to modify the surfaces of magnesium and its alloys in aspect of both microstructural features and compositional alterations in order to impart desirable qualities. Laser surface melting (LSM) has been widely utilized to modify the surfaces of magnesium and related alloys. The development of tiny dendritic grains (1–10 μm) in the changed region is principally responsible for the improvement in the surface characteristics. Additionally, LSM alters the composition of the laser-melted region due to the selective evaporation of elements such

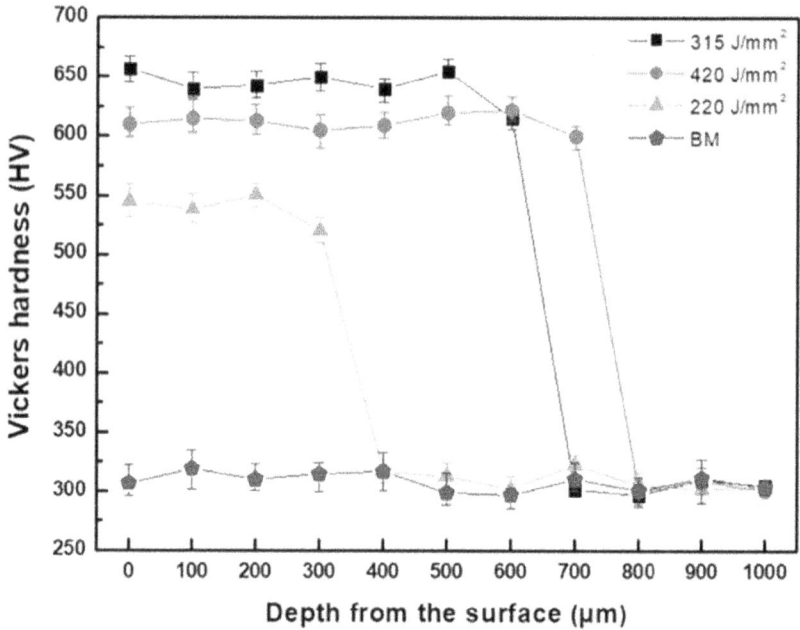

FIGURE 5.5 Vickers microhardness values of the base material and LS-engineered steel [87].

TABLE 5.1
The Wear and Corrosion Data of the Laser-Surface-Treated P20-Improved Steels and Base Metals at Varying Laser Densities

	Wear Loss ($\times 10^{-3}$ mm³) for Load 10N	Corrosion Potential (V/SCE)
Base metal	7.06 ± 0.64	-0.967
315 J/mm²	3.98 ± 0.31	-0.937
420 J/mm²	4.06 ± 0.20	-0.985

---

as magnesium and zinc. The corrosion behaviour of alloys heated by laser often depends on such compositional changes [88].

Dube et al. [89] studied LSM of AZ91 and AM60B. Due to the rapid cooling rates reached during laser processing, LSM led to grain refinement and the creation of a fine network of dendrites. ACM720 alloy was the subject of an investigation by Mondal et al., [90] who discovered substantial grain refinement there. Compared to the as-cast alloy (40–135 μm), the laser-melted region had grains that were 1.5–7 μm in size. Figure 5.6a depicts the cross section of ACM720 alloy that has been laser-melted and demonstrates a defect-free interface. Figure 5.6b and c depicts the microstructure of the laser-treated area where refined grains were found in the cross section and top

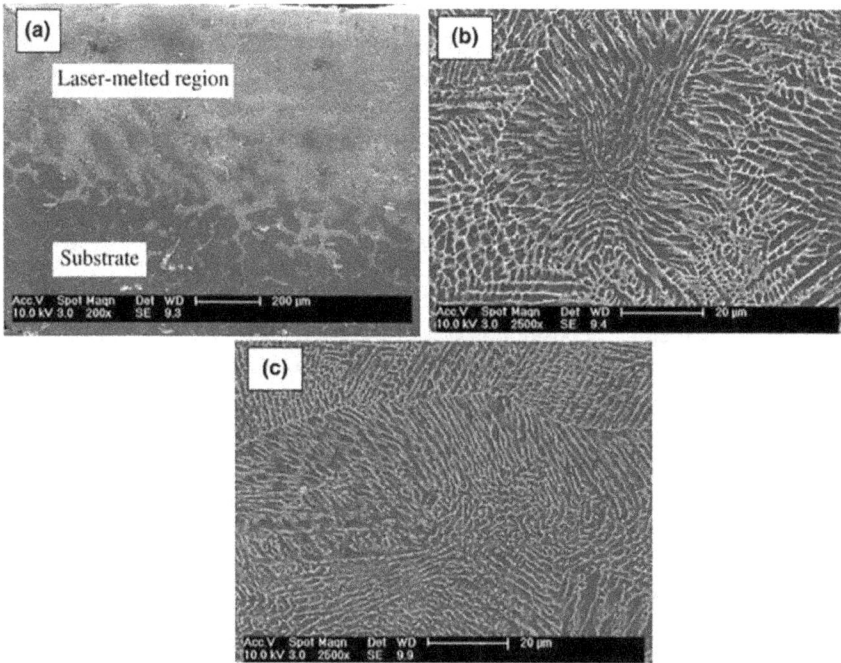

FIGURE 5.6   SEM micrographs from LSM: (a) cross section, (b) enlarged view of LM region, and (c) magnified view in top surface [90].

surface. Mixed cellular and dendritic grains made up the microstructure, which had a mean dendritic arm spacing of 2 μm. The surface of the laser-melted material contained phases including α-Mg, Al_8Mn_5, and $Ca_{31}Sn_{20}$, according to an XRD examination. The Al_2Ca phase, which was present in the alloy as-cast, was dissolved as a result of the rapid melting and solidification that occurred during LSM.

The wear resistance of laser-melted magnesium alloys improves together with an increase in surface hardness. Majumdar et al. [91] also explored at the wear characteristics of MEZ alloys that have been laser-surface-melted. Due to the increase in microhardness and grain refinement brought forth by LSM, wear resistance was enhanced. Abbas et al. [92,93] examined the wear resistance of AZ31 and AZ61 that had been laser-surface-melted. According to a report, the production of β-$Mg_{17}Al_{12}$ made laser-surface-melted alloys more resistant to wear. Galun and Mordike investigated the corrosion behaviour of a laser-alloyed magnesium substrate with Al, Cu, Ni, Si, and a mixture of these metals. Al- and Ni-based LSA showed improved corrosion resistance. Ming et al. [94] explored at how laser-surface-alloyed AZ91D with Al-Si developed microstructure and responded to corrosion. The production of dispersed intermetallic complexes such as Mg_2Si, γ-$Al_{12}Mg_17$, and β-Al_3Mg_2 was the primary factor for the increase in corrosion resistance.

Ceramics and ceramic-reinforced composites (CCRCs) exhibit extraordinary qualities at high temperatures, strong wear resistance, and excellent chemical inertness.

These qualities make them excellent for use in the high-end technical sectors of bio-medicine, aerospace, electronics, and others. Due to the aforementioned characteristics, they are expensive and energy-intensive to produce using traditional production techniques. Triantafyllidis et al. [95] conducted various studies on the application of laser surface treatment to refractory ceramics, primarily those based on Al_2O_3. Further research revealed that ceramic surfaces with no cracks had superior characteristics and these properties were required for corrosion resistance applications. Ester et al., also have similar findings. The authors concluded that laser-treated area of size 50 mm×7 mm^2 has a crack-free region. However, recently for CCRC coatings, laser additive manufacturing technology with high-power laser beams is being explored. Laser melting deposition is one of the developed laser additive manufacturing processes that has been successfully applied to coating surfaces with CCRC layers [96].

By adjusting the ZrO_2 concentration and powder preparation settings, ZrO_2 (95 vol%)+Al_2O_3 (5.0 vol%) exhibit enhanced toughness, exceptional corrosion and heat resistances, respectable biocompatibility, and good mechanical properties [97–99]. The developed Al_2O_3+ZrO_2 eutectic ceramic has an average microhardness of 17.15 GPa and a fracture toughness of 4.79 MPam$^{1/2}$. The microstructure of the developed coating, which consists of light and dark phases, is depicted in Figure 5.7.

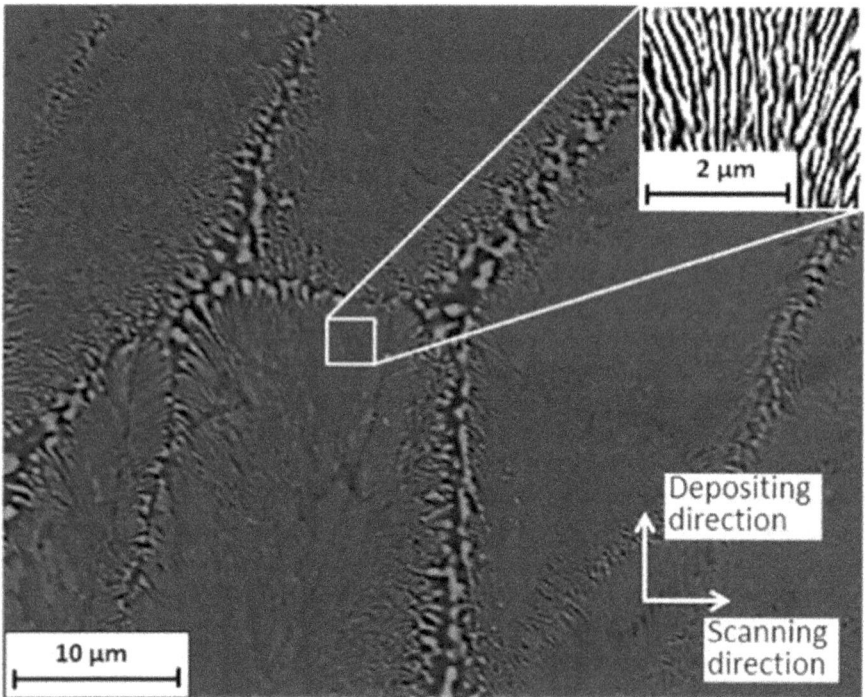

FIGURE 5.7   Microstructure of the Al_2O_3+ZrO_2 ceramic [96].

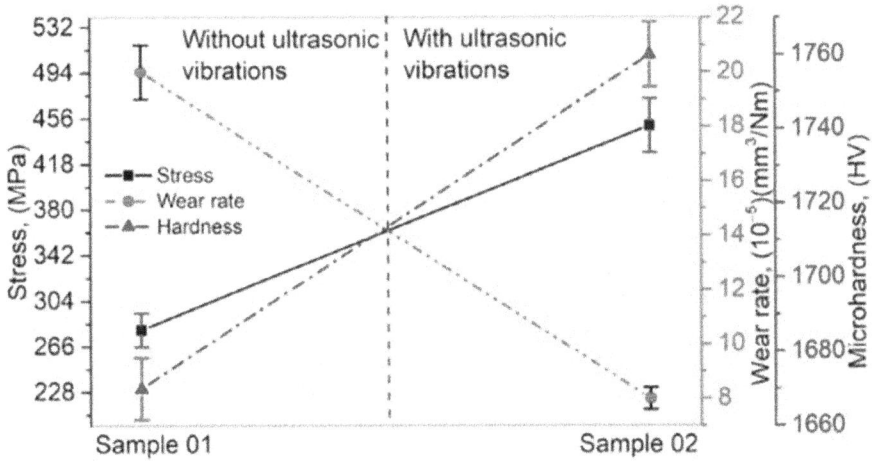

FIGURE 5.8   Stress, microhardness, and wear rate of the developed $ZrO_2 + Al_2O_3$ coatings on Tibase metal: with and without use of ultrasonic vibrations [96].

ZrO_2 is the tetragonal light phase, whereas Al_2O_3 is the dark phase. The $Al_2O_3 + ZrO_2$ eutectic ceramic's main component is its lamellar microstructure, which also plays a crucial role in determining the specimens' mechanical characteristics. By including an ultrasonic vibration in the laser melting deposition process, these microstructures were further improved and uniformized, comparable to 60.29 nm eutectic spacing [100]. The results of ultrasonic vibrations are depicted in Figure 5.8. It was discovered that the deposited coatings' stress-bearing ability (280–450 MPa) and hardness (1670–1760 HV) have both increased, while the wear rate ($20 \times 10^{-5} - 8 \times 10^{-5}$ mm³/Nm) has substantially lowered. This is due to the fact that the laser melting deposition process in combination with ultrasonic vibrations causes the homogenized material dispersion and grain refinement, which enhance the deposited materials' characteristics.

The laser melting deposition technique was used to deposit the coatings of $CaF_2 + Al_2O_3$ ceramics on an aluminium oxide substrate [101]. The CaF_2 particles were evenly distributed throughout the Al_2O_3 matrix, according to examinations done with a SEM. With its substantially lower friction coefficient, CaF_2, coatings that are both self-lubricating and resistant to wear became much better.

The laser melting deposition method was used to provide a coating of tungsten carbide (WC) + cobalt (Co) on a steel substrate to safeguard heat exchanger tubes against issues of erosion and corrosion [102]. Results show that when compared to pure material, the deposited coatings provided good hardness and acceptable wear resistance. The laser melting deposition technique was used to deposit Ni + TiC coatings on low carbon steel [103]. Analysis was done on the effects of TiC volume percent on hardness, phase transformation, wear resistance, and microstructure evolution. The wear resistance value changes from 20 to 6×10^{-3} g, while the hardness increased from 365.6 to 1897.6 HV.

Due to their high strength-to-weight ratio, superior corrosion resistance, and exceptional biocompatibility, **Ti and its alloys** are candidate materials for use in the aerospace and biomedical areas [104–119]. The useful service life of the parts manufactured from Ti and its alloys is, however, reduced due to insufficient wear and corrosion. These attributes may be increased by appropriately modifying the surface microstructures. When it comes to changing the surface area of reactive metals like titanium, LSE is extremely important. Biswas et al. [120] with aim to increase the wear resistance of Ti-6Al-4V implant treated it by LSM and nitriding. After this, the laser surface treatment's effects on corrosion resistance in a simulated body fluid and biocompatibility will be investigated. In contrast to the 260 Hv of the substrate as received, the microhardness may be increased up to 450 Hv in the case of LSM and 900–950 Hv in the event of laser surface nitriding. The microstructure was significantly refined as a result of LSM, and acicular martensite with partial suppression of the β-phase was formed. On the other side, TiN dendrites formed in the α-Ti matrix as a result of laser surface nitriding. In comparison with Ti-6Al-4V as-received, surface melting greatly raises the corrosion potential (Ecorr) and main potential for pit formation.

Yue et al. [121] conducted excimer laser treatment of alloy Ti-6Al-4V with motive to enhance its corrosion resistance. Under two separate gas conditions, argon and nitrogen, laser surface treatment was carried out. A pretty smooth and crack-free modulated surface layer was created by excimer laser surface treatment. The Ti alloy's corrosion resistance was dramatically increased by excimer laser surface treatment.

5.4 MODERN-DAY APPLICATIONS OF LSE

LSE can be used to target particular regions of a work item due to the directional aspect of laser light. However, compared to other technologies, the main drawbacks of LSE are its inability to be applied to large areas and its expensive initial equipment costs. LSE, on the other hand, is a flexible and adaptable approach that is environmentally benign. Compared to other surface modification techniques, LSE has a number of advantages, including being precise, selective, and ultra-fast as well as offering access to a variety of geometrical shapes. In this section, few modern-day applications of LSE were explored.

5.4.1 STENTS

Due to their small size and need for accurate, individualized, and clean modification, stents lend themselves particularly well to LSE. In order to achieve various textures and surface chemistries on stents or materials related to stents, several research teams have researched LSE, assessing the resulting stent performance in vitro and/or in vivo. A viable method for creating pillars, pits, grooves, and intricate freeform surface structures is laser texturing. Due to their simplicity of production, grooved surface structures are still widely used for laser surface modification of stents. By laser ablation, Li et al. [122] created groove-like patterns on 316L stents. They found that by applying optimized laser parameters, it is possible to create a biomimetic grooved structure that resembles the luminal surface of a rat aorta.

5.4.2 Steering Gear Assembly Hardening

The hardening of steering gear assemblies is the first industrial use of laser heat treatment by Ready and Farson Dave [123]. Automobile power steering pumps were initially intended to be heated in a batch furnace, but as energy costs rose, a 24-hour furnace schedule became too expensive. In order to prevent expected wear from the piston operating under increased loads, engineers at a U.S. car company's steering gear division chose to create a localized wear-resistant surface on the inner bore of the pump housing rather than heating 5 kg of ferritic-malleable cast iron. Five hardened tracks judiciously arranged would create enough wear resistance without causing part distortion, according to CO_2 laser experimentation. In actual use, a spinning optic device directs the beam of a CO_2 laser with a kilowatt output to the inside bore of the pump housing.

5.4.3 Turbine Blades Repair by Laser Cladding

Political and environmental considerations make reliable and efficient electricity generating a crucial global issue. The cost of bringing machines offline for service and removing damaged blades for repair makes it expensive. To overcome this issue, in situ laser cladding of turbine blades was demonstrated worldwide first time by Brandt et al. [124] in 2004. When the turbine was examined in May 2005, the laser-repaired blades performed well and showed no signs of deterioration. By using more repaired blades and better processing, the October 2005 study significantly boosted confidence in the method. In order to commercialize the invention, Hardwear Pty Ltd was established. In 2007 and 2008, the company successfully completed two commercial contracts.

5.4.4 Laser Cladding of Truck Wheel Spindles

In Australia's mining industry, big dump trucks (CAT777) are frequently used. Wheel spindles are a high-wear, high-cost component. A growing number of dump trucks are "parked up" in anticipation of maintenance overhauls as a result of the mining downturn. Waiting for new or replacement parts can save a lot of time and money if these and other expensive components are reclaimed using laser cladding. The same laser cladding method was used by *LaserBond* to repair these worn-out or damaged surfaces. Because of the metallurgical bond, applied layers can be employed in high-impact, substantially loaded conditions without being at risk of spalling or overlay separation. The laser cladding energy's limitless controllability enables the reduction of unwelcome thermal disintegration of hard phases like carbides, which causes substrate dilution, decomposition, and deformation effects common to alternative restoration techniques.

5.5 CHALLENGES DURING LSE

It has been successfully proven that using laser-based methods can increase wear, corrosion, and fatigue resistance. However, because of the intricacy of heat transfer rate and fluid flow effects during process, fabricating consistently defect-free surfaces remains difficult. Current knowledge of effects like selective vaporization of alloying elements, penetration of second-phase particles, and the development of unwanted residual stresses is required to be improved in order to create environmentally friendly laser-based surface modification techniques for light alloys.

For light alloys, LSM is the most widely utilized LSE technique. The majority of property enhancements result from microstructure refinement up to a depth of around 1 mm. For instance, LSM frequently produces refined primary α-Mg dendritic grains and fully/partially separated eutectic phases in as-cast Mg alloys. Although microstructure refinement brought about by LSM typically improves surface hardness and wear resistance, findings on improved corrosion resistance are inconsistent. The selective evaporation of alloying materials during LSM procedures is one of the main problems. For the Mg alloys, for instance, it has been seen that Mg and Zn are selectively vaporized. When corrosion resistance improvement is crucial, LSA techniques might be more appropriate than LSM. To alter the surface composition in LSA, a selected alloying element having desirable properties is incorporated to the molten pool. Intermetallic phases are frequently formed as a result of LSA, which simultaneously improves wear resistance and corrosion resistance. Laser surface nitriding is another approach to improve surface properties. The controlled nitrogen gas is used to laser-melt the alloy's surface. Due to the development of thin metal nitrides, the microhardness and surface properties of the nitrided surfaces dramatically increase.

Through the use of laser composite surfacing (LCS), which incorporates hard ceramic particles into the laser-melted surface, it is possible to create efficient tribological surfaces for lighter alloys. To achieve uniform dispersion of reinforced hard ceramic particles in alloy matrices, controlling of parameters (for laser and powder injection) becomes crucial. With LCS, it is common to see particle agglomeration at the surface, a lack of hard particle penetration into the melt, undesired interfacial reactions, and a lack of particle wetting. These flaws are easily capable of causing particles to separate from the surface and to trigger three-body abrasive wear mechanics. Additionally, several attempts have been made to introduce hard-phase nanoparticles into the alloy matrixes, but LCS appears to be restricted to comparatively coarser particle reinforcement, i.e., greater than 50 μm. The highest depth of penetration, even when using these coarser particles, is frequently just under 1/2 of the maximum depth of laser melting. In order to prevent the total melting of the hard particles, laser processing settings must also be tuned. In some cases, preplacing the powder on the substrate rather than injecting it directly into the melt during LCS can result in a better dispersion of hard particles in the alloy matrix. A similar procedure is LSC, which involves laser deposition (bonding/fusing) of a reasonably thick layer of any superior material/composite on the substrate (thickness >500 μm). Care must be taken to avoid interfacial cracking since the material deployed for cladding and the material of substrate vary in chemical compositions.

Laser shock peening (LSP) is a fairly recent method for surface engineering light alloys. Here, high-pressure plasma is produced at the treated surface by means of brief laser pulses. The ensuing shock waves result in plastic deformation and add residual stresses (compressive in nature) in the surface. The microstructure and mechanical characteristics of the treated metal are positively impacted by the laser-induced compressive residual stress field. The mechanical performance is enhanced by microstructural alterations like dynamic recrystallization and grain refinement. For instance, the material's in-service application may result in a delay in the onset of surface cracks, and fatigue life may be greatly increased. Additionally, LSP can improve the treated metal's resistance to oxidation and corrosion. In order to effectively use the energy of the shock wave produced by the laser to induce compressive surface stresses, the substrate's surface needs to be subjected to almost entirely mechanical stresses [125]. But tensile residual stresses emerge if the exposed surface of the laser-treated substrate is overheated during laser irradiation [126]. And it should be taken care off. Sundar et al. [127] covered the surface with an appropriate sacrificial layer (thermo-absorptive in nature) to prevent thermal exposure during treatment to the substrate's surface to restrict the outcome of tensile residual stresses.

There has been an increase in interest to improve the efficiency of biodegradable materials, which were used for stents in recent years, but no systematic evaluation of the impact of wavelength, frequency, and pulse width on the degradation process, particularly for polymeric biodegradable materials, has been done. There exist still many obstacles to simultaneously increasing strength and controlling degradation, which presents prospects for future research in this area.

5.6   CONCLUSION AND FUTURE PERSPECTIVE

These laser-based innovations are anticipated to play a significant role in surface engineering, as there is little question that significantly superior surface qualities are required for newer applications than those in the past. Interest in LSE is anticipated to persist and advancements in this field are going on as long as attempts to achieve energy independence and environmental protection are made. Due to their better material properties, i.e., low densities and relatively high strengths, lightweight materials—most notably those based on Al, Mg, and Ti—have long been studied for a variety of applications. Expanding the utility of light alloys in new applications is becoming more popular. LSE nowadays is widely used in biomedical area. For instance, with the use of LSE, it is possible to treat stent materials and create a variety of 3D textures, such as pillars, pores, nanowires, and freeform architectures.

In near future, LSE for stent applications will need to take into account both surface chemical design and innovative texturing. As an illustration, Cui et al. [128,129] developed a creative laser texturing design. The combo of circle and linear scanning patterns improved NiTi's resistance to corrosion and boosted osteoblast cells' bioactivity. Future study should pay special consideration to these unique designs because, although they have not yet been created for stent surface texturing, these 3D freeform textures may enhance stent performance owing to their nano- and micrometric properties.

The use of laser surface texturing technology for metal surface treatment (micromachining, functionalization, and drilling) in numerous applications is gaining popularity. The employment of lasers as a technique for metal structure surface preparation before organic coating application is not well documented in the literature and can be explored in near future. Additionally, nothing is known about how the laser surface treatment may affect these coatings' ability to resist corrosion in very aggressive conditions. Recently, Caraguay et al., [56] employed nanosecond laser surface texturing technique to develop grid and groove-like surface patterns. The authors reported that for the purpose of preparing steel surfaces deployed in offshore maritime conditions, nanosecond laser surface texturing looks to be an effective technique. This technique promotes mechanical interlocking within the deposited coating due to the structure of grooves. In nut shell, laser technology is projected to become more appealing in future as a result of advancements combined with creative concepts (i.e., the use of extra heat sources and ultrasonic vibrations) since they allow for more flexible manufacturing and better product quality at a substantially lower price.

REFERENCES

1. Prashar, G., Vasudev, H., Thakur, L.: High-temperature oxidation and erosion resistance of Ni-based thermally-sprayed coatings used in power generation machinery: A review. *Surf. Rev. Lett.* 29, 2230003 (2022). https://doi.org/10.1142/S0218625X22300039.
2. Vasudev, H., Prashar, G., Thakur, L., Bansal, A.: Electrochemical corrosion behavior and microstructural characterization of HVOF sprayed Inconel718-Al$_2$O$_3$ composite coatings. *Surf. Rev. Lett.* 29, 2250017 (2022). https://doi.org/10.1142/S0218625X22500172.
3. Prashar, G., Vasudev, H.: A review on the influence of process parameters and heat treatment on the corrosion performance of Ni-based thermal spray coatings. *Surf. Rev. Lett.* 29, 1–18 (2022). https://doi.org/10.1142/S0218625X22300015.
4. Singh, G., Vasudev, H., Bansal, A., Vardhan, S.: Influence of heat treatment on the microstructure and corrosion properties of the Inconel-625 clad deposited by microwave heating. *Surf. Topogr. Metrol. Prop.* 9, 025019 (2021). https://doi.org/10.1088/2051-672X/abfc61.
5. Vasudev, H., Prashar, G., Thakur, L., Bansal, A.: Microstructural characterization and electrochemical corrosion behaviour of HVOF sprayed Alloy718-nanoAl$_2$O$_3$ composite coatings. *Surf. Topogr. Metrol. Prop.* 9, 035003 (2021). https://doi.org/10.1088/2051-672X/ac1044.
6. Singh, P., Bansal, A., Vasudev, H., Singh, P.: In situ surface modification of stainless steel with hydroxyapatite using microwave heating. *Surf. Topogr. Metrol. Prop.* 9, 035053 (2021). https://doi.org/10.1088/2051-672X/ac28a9.
7. Dutta Majumdar, J., Manna, I.: Laser surface engineering of titanium and its alloys for improved wear, corrosion and high-temperature oxidation resistance. In: *Laser Surface Engineering.* pp. 483–521. Elsevier (2015). https://doi.org/10.1016/B978-1-78242-074-3.00021-0.
8. Vasudev, H., Prashar, G., Thakur, L., Bansal, A.: Electrochemical corrosion behavior and microstructural characterization of HVOF sprayed Inconel-718 coating on gray cast iron. *J. Fail. Anal. Prev.* 21, 250–260 (2021). https://doi.org/10.1007/s11668-020-01057-8.
9. Vasudev, H., Thakur, L., Singh, H., Bansal, A.: A study on processing and hot corrosion behaviour of HVOF sprayed Inconel718-nano Al$_2$O$_3$ coatings. *Mater. Today Commun.* 25, 101626 (2020). https://doi.org/10.1016/j.mtcomm.2020.101626.
10. Singh, J., Kumar, S., Mohapatra, S.K.: Erosion wear performance of Ni-Cr-O and NiCrBSiFe-WC(Co) composite coatings deposited by HVOF technique. *Ind. Lubr. Tribol.* 71, 610–619 (2019). https://doi.org/10.1108/ILT-04-2018-0149.
11. Singh, J., Kumar, S., Mohapatra, S.K.: Erosion tribo-performance of HVOF deposited Stellite-6 and Colmonoy-88 micron layers on SS-316L. *Tribol. Int.* 147, 105262 (2020). https://doi.org/10.1016/j.triboint.2018.06.004.
12. Singh, J., Singh, S., Gill, R.: Applications of biopolymer coatings in biomedical engineering. *J. Electrochem. Sci. Eng.* (2022). https://doi.org/10.5599/jese.1460.

13. Singh, J.: Tribo-performance analysis of HVOF sprayed 86WC-10Co4Cr & Ni-Cr$_2$O$_3$ on AISI 316L steel using DOE-ANN methodology. *Ind. Lubr. Tribol.* 73, 727–735 (2021). https://doi.org/10.1108/ILT-04-2020-0147.

14. Singh, J.: Application of thermal spray coatings for protection against erosion, abrasion, and corrosion in hydropower plants and offshore industry. In: Thakur, L. and Vasudev, H. (eds.) *Thermal Spray Coatings.* pp. 243–283. CRC Press, Boca Raton (2021). https://doi.org/10.1201/9781003213185-10.

15. Singh, J.: Slurry erosion performance analysis and characterization of high-velocity oxy-fuel sprayed Ni and Co hardsurfacing alloy coatings. *J. King Saud Univ. - Eng. Sci.* (2021). https://doi.org/10.1016/j.jksues.2021.06.009.

16. Kumar, D., Yadav, R., Singh, J.: Evolution and adoption of microwave claddings in modern engineering applications. In: *Advances in Microwave Processing for Engineering Materials.* pp. 134–153. CRC Press, Boca Raton (2022). https://doi.org/10.1201/9781003248743-8.

17. Dutta Majumdar, J., Manna, I.: Laser material processing. *Int. Mater. Rev.* 56, 341–388 (2011). https://doi.org/10.1179/1743280411Y.0000000003.

18. Singh, J., Kumar, S., Mohapatra, S.: Study on role of particle shape in erosion wear of austenitic steel using image processing analysis technique. *Proc. Inst. Mech. Eng. Part J J. Eng. Tribol.* 233, 712–725 (2019). https://doi.org/10.1177/1350650118794698.

19. Singh, J., Kumar, S., Mohapatra, S.K.: Tribological analysis of WC–10Co–4Cr and Ni–20Cr$_2$O$_3$ coating on stainless steel 304. *Wear.* 376–377, 1105–1111 (2017). https://doi.org/10.1016/j.wear.2017.01.032.

20. Kumar, K., Kumar, S., Singh, G., Singh, J., Singh, J.: Erosion wear investigation of HVOF sprayed WC-10Co4Cr coating on slurry pipeline materials. *Coatings.* 7, 54 (2017). https://doi.org/10.3390/coatings7040054.

21. Singh, J.: Analysis on suitability of HVOF sprayed Ni-20Al, Ni-20Cr and Al-20Ti coatings in coal-ash slurry conditions using artificial neural network model. *Ind. Lubr. Tribol.* 71, 972–982 (2019). https://doi.org/10.1108/ILT-12-2018-0460.

22. Singh, J., Kumar, S., Singh, J.P., Kumar, P., Mohapatra, S.K.: CFD modeling of erosion wear in pipe bend for the flow of bottom ash suspension. *Part. Sci. Technol.* 37, 275–285 (2019). https://doi.org/10.1080/02726351.2017.1364816.

23. Singh, J., Kumar, S., Mohapatra, S.K.: Tribological performance of Yttrium (III) and Zirconium (IV) ceramics reinforced WC–10Co4Cr cermet powder HVOF thermally sprayed on X2CrNiMo-17-12-2 steel. *Ceram. Int.* 45, 23126–23142 (2019). https://doi.org/10.1016/j.ceramint.2019.08.007.

24. Singh, J., Kumar, S., Mohapatra, S.K.: An erosion and corrosion study on thermally sprayed WC-Co-Cr powder synergized with Mo$_2$C/Y$_2$O$_3$/ZrO$_2$ feedstock powders. *Wear.* 438–439, 102751 (2019). https://doi.org/10.1016/j.wear.2019.01.082.

25. Singh, J., Singh, J.P.: Numerical analysis on solid particle erosion in elbow of a slurry conveying circuit. *J. Pipeline Syst. Eng. Pract.* 12, 04020070 (2021). https://doi.org/10.1061/(asce)ps.1949-1204.0000518.

26. Singh, J., Singh, C.: Numerical analysis of heat dissipation from a heated vertical cylinder by natural convection. *Proc. Inst. Mech. Eng. Part E J. Process Mech. Eng.* 231, 405–413 (2017). https://doi.org/10.1177/0954408915600109.

27. Singh, J.: A review on mechanisms and testing of wear in slurry pumps, pipeline circuits and hydraulic turbines. *J. Tribol.* 143, 1–83 (2021). https://doi.org/10.1115/1.4050977.

28. Singh, J., Mohapatra, S.K., Kumar, S.: Performance analysis of pump materials employed in bottom ash slurry erosion conditions. *J. Tribol.* 30, 73–89 (2021).

29. Singh, J., Singh, S.: Neural network prediction of slurry erosion of heavy-duty pump impeller/casing materials 18Cr-8Ni, 16Cr-10Ni-2Mo, super duplex 24Cr-6Ni-3Mo-N, and grey cast iron. *Wear.* 476, 203741 (2021). https://doi.org/10.1016/j.wear.2021.203741.

30. Singh, J.: Wear performance analysis and characterization of HVOF deposited Ni–20Cr$_2$O$_3$, Ni–30Al$_2$O$_3$, and Al$_2$O$_3$–13TiO$_2$ coatings. *Appl. Surf. Sci. Adv.* 6, 100161 (2021). https://doi.org/10.1016/j.apsadv.2021.100161.
31. Singh, J., Singh, S., Pal Singh, J.: Investigation on wall thickness reduction of hydropower pipeline underwent to erosion-corrosion process. *Eng. Fail. Anal.* 127, 105504 (2021). https://doi.org/10.1016/j.engfailanal.2021.105504.
32. Kumar, P., Kumar, S., Singh, J.: Rheological and computational analysis of crude oil transportation. *Int. J. Mech. Aerospace, Ind. Mechatron. Manuf. Eng.* 11, 429–432 (2017).
33. Singh, J., Singh, S.: Neural network supported study on erosive wear performance analysis of Y$_2$O$_3$/WC-10Co4Cr HVOF coating. *J. King Saud Univ. - Eng. Sci.* (2022). https://doi.org/10.1016/j.jksues.2021.12.005.
34. Singh, J., Singh, J.P.: Performance analysis of erosion resistant Mo$_2$C reinforced WC-CoCr coating for pump impeller with Taguchi's method. *Ind. Lubr. Tribol.* 74, 431–441 (2022). https://doi.org/10.1108/ILT-05-2020-0155.
35. Singh, S., Garg, J., Singh, P., Singh, G., Kumar, K.: Effect of hard faced Cr-alloy on abrasive wear of low carbon rotavator blades using design of experiments. *Mater. Today Proc.* 5, 3390–3395 (2018). https://doi.org/10.1016/j.matpr.2017.11.583.
36. Singh, J., Kumar, S., Mohapatra, S.K.: Optimization of erosion wear influencing parameters of HVOF sprayed pumping material for coal-water slurry. *Mater. Today Proc.* 5, 23789–23795 (2018). https://doi.org/10.1016/j.matpr.2018.10.170.
37. Singh, J., Kumar, S., Singh, G.: Taguchi's approach for optimization of tribo-resistance parameters Forss304. *Mater. Today Proc.* 5, 5031–5038 (2018). https://doi.org/10.1016/j.matpr.2017.12.081.
38. Singh, J., Kumar, M., Kumar, S., Mohapatra, S.K.: Properties of glass-fiber hybrid composites: A review. *Polym. Plast. Technol. Eng.* 56, 455–469 (2017). https://doi.org/10.1080/03602559.2016.1233271.
39. Singh, J., Kumar, S., Mohapatra, S.K., Kumar, S.: Shape simulation of solid particles by digital interpretations of scanning electron micrographs using IPA technique. *Mater. Today Proc.* 5, 17786–17791 (2018). https://doi.org/10.1016/j.matpr.2018.06.103.
40. Kumar, P., Kumar, S., Singh, J.: Effect of natural surfactant on the rheological characteristics of heavy crude oil. *Mater. Today Proc.* 5, 23881–23887 (2018). https://doi.org/10.1016/j.matpr.2018.10.180.
41. Kumar, P., Singh, J.: Computational study on effect of Mahua natural surfactant on the flow properties of heavy crude oil in a 90° bend. *Mater. Today Proc.* 43, 682–688 (2021). https://doi.org/10.1016/j.matpr.2020.12.612.
42. Singh, J., Kumar, S., Mohapatra, S.K.: Study on solid particle erosion of pump materials by fly ash slurry using Taguchi's orthogonal array. *Tribol. - Finnish J. Tribol.* 38, 31–38 (2021). https://doi.org/10.30678/fjt.97530.
43. Kumar, S., Singh, J., Mohapatra, S.K.: Influence of particle size on leaching characteristic of fly ash. In: *15th International Conference on Environmental Science and Technology, 31 August to 2 September 2017*. p. 01243. Rhodes, Greece (2017).
44. Singh, J., Singh, C.: Computational analysis of convective heat transfer across a vertical tube. *FME Trans.* 49, 932–940 (2021). https://doi.org/10.5937/fme2104932S.
45. Singh, J., Singh, S.: A review on machine learning aspect in physics and mechanics of glasses. *Mater. Sci. Eng. B.* 284, 115858 (2022). https://doi.org/10.1016/j.mseb.2022.115858.
46. Kumar, S., Singh, J., Mohapatra, S.K.: Role of particle size in assessment of physicochemical properties and trace elements of Indian fly ash. *Waste Manag. Res.* 36, 1016–1022 (2018). https://doi.org/10.1177/0734242X18804033.
47. Kumar, P., Singh, J., Singh, S.: Neural network supported flow characteristics analysis of heavy sour crude oil emulsified by ecofriendly bio-surfactant utilized as a replacement of sweet crude oil. *Chem. Eng. J. Adv.* 11, 100342 (2022). https://doi.org/10.1016/j.ceja.2022.100342.

48. Singh, J., Singh, J.P., Singh, M., Szala, M.: Computational analysis of solid particle-erosion produced by bottom ash slurry in 90° elbow. In *MATEC Web of Conferences*, vol. 252, p. 04008. EDP Sciences (2019). https://doi.org/0.1051/matecconf/201925204008.

49. Singh, J.: *Investigation on Slurry Erosion of Different Pumping Materials and Coatings.* Thapar Institute of Engineering and Technology, Patiala, India (2019).

50. Singh, J., Singh, R., Singh, S., Vasudev, H., Kumar, S.: Reducing scrap due to missed operations and machining defects in 90PS pistons. *Int. J. Interact. Des. Manuf.* (2022). https://doi.org/10.1007/s12008-022-01071-0.

51. Singh, J., Singh, C.: Study of buoyant force acting on different fluids moving over a horizontal plate due to forced convection. *IOP Conf. Ser. Mater. Sci. Eng.* 710, 012044 (2019). https://doi.org/10.1088/1757-899X/710/1/012044.

52. Kumar, S., Singh, M., Singh, J., Singh, J.P., Kumar, S.: Rheological characteristics of uni/bi-variant particulate iron ore slurry: Artificial neural network approach. *J. Min. Sci.* 55, 201–212 (2019). https://doi.org/10.1134/S1062739119025468.

53. Singh, R., Toseef, M., Kumar, J., Singh, J.: Benefits and challenges in additive manufacturing and its applications. In: Kaushal, S., Singh, I., Singh, S., and Gupta, A. (eds.) *Sustainable Advanced Manufacturing and Materials Processing.* pp. 137–157. CRC Press, Boca Raton (2022). https://doi.org/10.1201/9781003269298-8.

54. Singh, J., Singh, S., Verma, A.: Artificial intelligence in use of ZrO_2 material in biomedical science. *J. Electrochem. Sci. Eng.* (2022). https://doi.org/10.5599/jese.1498.

55. Singh, J., Gill, H.S., Vasudev, H.: Computational fluid dynamics analysis on effect of particulate properties on erosive degradation of pipe bends. *Int. J. Interact. Des. Manuf.* (2022). https://doi.org/10.1007/s12008-022-01094-7.

56. Caraguay, S.J., Pereira, T.S., Giacomelli, R.O., Cunha, A., Pereira, M., Xavier, F.A.: The effect of laser surface textures on the corrosion resistance of epoxy coated steel exposed to aggressive environments for offshore applications. *Surf. Coatings Technol.* 437, 128371 (2022). https://doi.org/10.1016/j.surfcoat.2022.128371.

57. Vasudev, H., Thakur, L., Singh, H., Bansal, A.: An investigation on oxidation behaviour of high velocity oxy-fuel sprayed Inconel718-Al_2O_3 composite coatings. *Surf. Coatings Technol.* 393, 125770 (2020). https://doi.org/10.1016/j.surfcoat.2020.125770.

58. Vasudev, H., Thakur, L., Bansal, A., Singh, H., Zafar, S.: High temperature oxidation and erosion behaviour of HVOF sprayed bi-layer Alloy-718/NiCrAlY coating. *Surf. Coatings Technol.* 362, 366–380 (2019). https://doi.org/10.1016/j.surfcoat.2019.02.012.

59. Prashar, G., Vasudev, H.: Structure-property correlation and high-temperature erosion performance of Inconel625-Al_2O_3 plasma-sprayed bimodal composite coatings. *Surf. Coatings Technol.* 439, 128450 (2022). https://doi.org/10.1016/j.surfcoat.2022.128450.

60. Singh, M., Vasudev, H., Singh, M.: Surface protection of SS-316L with boron nitride based thin films using radio frequency magnetron sputtering technique. *J. Electrochem. Sci. Eng.* 25, 851–863 (2022). https://doi.org/10.5599/jese.1247.

61. Singh, G., Vasudev, H., Bansal, A., Vardhan, S., Sharma, S.: Microwave cladding of Inconel-625 on mild steel substrate for corrosion protection. *Mater. Res. Express.* 7, 026512 (2020). https://doi.org/10.1088/2053-1591/ab6fa3.

62. Singh, M., Vasudev, H., Kumar, R.: Corrosion and tribological behaviour of BN thin films deposited using magnetron sputtering. *Int. J. Surf. Eng. Interdiscip. Mater. Sci.* 9, 24–39 (2021). https://doi.org/10.4018/IJSEIMS.2021070102.

63. Vasudev, H., Thakur, L., Singh, H., Bansal, A.: Erosion behaviour of HVOF sprayed Alloy718-nano Al_2O_3 composite coatings on grey cast iron at elevated temperature conditions. *Surf. Topogr. Metrol. Prop.* 9, 035022 (2021). https://doi.org/10.1088/2051-672X/ac1c80.

64. Bansal, A., Vasudev, H., Sharma, A.K., Kumar, P.: Investigation on the effect of post weld heat treatment on microwave joining of the Alloy-718 weldment. *Mater. Res. Express.* 6, 086554 (2019). https://doi.org/10.1088/2053-1591/ab1d9a.

65. Vasudev, H., Thakur, L., Singh, H., Bansal, A.: Effect of addition of Al_2O_3 on the high-temperature solid particle erosion behaviour of HVOF sprayed Inconel-718 coatings. *Mater. Today Commun.* 30, 103017 (2022). https://doi.org/10.1016/j.mtcomm.2021.103017.

66. Vasudev, H.: Wear characteristics of Ni-WC powder deposited by using a microwave route on mild steel. *Int. J. Surf. Eng. Interdiscip. Mater. Sci.* 8, 44–54 (2020). https://doi.org/10.4018/IJSEIMS.2020010104.

67. Prashar, G., Vasudev, H., Bhuddhi, D.: Additive manufacturing: Expanding 3D printing horizon in industry 4.0. *Int. J. Interact. Des. Manuf.* (2022). https://doi.org/10.1007/s12008-022-00956-4.

68. Vasudev, H., Thakur, L., Singh, H., Bansal, A.: Mechanical and microstructural behaviour of wear resistant coatings on cast iron lathe machine beds and slides. *Met. Mater.* 56, 55–63 (2018). https://doi.org/10.4149/km_2018_1_55.

69. Prashar, G., Vasudev, H.: Structure–property correlation of plasma-sprayed Inconel625-Al_2O_3 bimodal composite coatings for high-temperature oxidation protection. *J. Therm. Spray Technol.* 31, 2385–2408 (2022). https://doi.org/10.1007/s11666-022-01466-1.

70. Prashar, G., Vasudev, H.: A comprehensive review on sustainable cold spray additive manufacturing: State of the art, challenges and future challenges. *J. Clean. Prod.* 310, 127606 (2021). https://doi.org/10.1016/j.jclepro.2021.127606.

71. Prashar, G., Vasudev, H., Thakur, L.: Performance of different coating materials against slurry erosion failure in hydrodynamic turbines: A review. *Eng. Fail. Anal.* 115, 104622 (2020). https://doi.org/10.1016/j.engfailanal.2020.104622.

72. Vasudev, H., Singh, G., Bansal, A., Vardhan, S., Thakur, L.: Microwave heating and its applications in surface engineering: A review. *Mater. Res. Express.* 6, 102001 (2019). https://doi.org/10.1088/2053-1591/ab3674.

73. Prashar, G., Vasudev, H.: High temperature erosion behavior of plasma sprayed Al_2O_3 coating on AISI-304 stainless steel. *World J. Eng.* 18, 760–766 (2021). https://doi.org/10.1108/WJE-10-2020-0476.

74. Vasudev, H., Singh, P., Thakur, L., Bansal, A.: Mechanical and microstructural characterization of microwave post processed Alloy-718 coating. *Mater. Res. Express.* 6, 1265f5 (2020). https://doi.org/10.1088/2053-1591/ab66fb.

75. Prashar, G., Vasudev, H.: Hot corrosion behavior of super alloys. *Mater. Today Proc.* 26, 1131–1135 (2020). https://doi.org/10.1016/j.matpr.2020.02.226.

76. Mehta, A., Vasudev, H., Singh, S.: Recent developments in the designing of deposition of thermal barrier coatings – A review. *Mater. Today Proc.* 26, 1336–1342 (2020). https://doi.org/10.1016/j.matpr.2020.02.271.

77. Singh, M., Vasudev, H., Kumar, R.: Microstructural characterization of BN thin films using RF magnetron sputtering method. *Mater. Today Proc.* 26, 2277–2282 (2020). https://doi.org/10.1016/j.matpr.2020.02.493.

78. Sunitha, K., Vasudev, H.: Microsrtructural and mechanical characterization of HVOF-sprayed Ni-based alloy coating. *Int. J. Surf. Eng. Interdiscip. Mater. Sci.* 10, 1–9 (2022). https://doi.org/10.4018/IJSEIMS.298705.

79. Ganesh Reddy Majji, B., Vasudev, H., Bansal, A.: A review on the oxidation and wear behavior of the thermally sprayed high-entropy alloys. *Mater. Today Proc.* 50, 1447–1451 (2022). https://doi.org/10.1016/j.matpr.2021.09.016.

80. Prashar, G., Vasudev, H.: Surface topology analysis of plasma sprayed Inconel625-Al_2O_3 composite coating. *Mater. Today Proc.* 50, 607–611 (2022). https://doi.org/10.1016/j.matpr.2021.03.090.

81. Satyavathi Yedida, V.V., Vasudev, H.: A review on the development of thermal barrier coatings by using thermal spray techniques. *Mater. Today Proc.* 50, 1458–1464 (2022). https://doi.org/10.1016/j.matpr.2021.09.018.

82. Singh, J., Vasudev, H., Singh, S.: Performance of different coating materials against high temperature oxidation in boiler tubes – A review. *Mater. Today Proc.* 26, 972–978 (2020). https://doi.org/10.1016/j.matpr.2020.01.156.
83. Sharma, Y., Singh, K.J., Vasudev, H.: Experimental studies on friction stir welding of aluminium alloys. *Mater. Today Proc.* 50, 2387–2391 (2022). https://doi.org/10.1016/j.matpr.2021.10.254.
84. Parkash, J., Saggu, H.S., Vasudev, H.: A short review on the performance of high velocity oxy-fuel coatings in boiler steel applications. *Mater. Today Proc.* 50, 1442–1446 (2022). https://doi.org/10.1016/j.matpr.2021.09.014.
85. Sunitha, K., Vasudev, H.: A short note on the various thermal spray coating processes and effect of post-treatment on Ni-based coatings. *Mater. Today Proc.* 50, 1452–1457 (2022). https://doi.org/10.1016/j.matpr.2021.09.017.
86. Dwivedi, D.K.: Surface engineering. In: *Surface Engineering.* Springer (2018). https://doi.org/10.1007/978-81-322-3779-2_1
87. Park, C., Sim, A., Ahn, S., Kang, H., Chun, E.-J.: Influence of laser surface engineering of AISI P20-improved mold steel on wear and corrosion behaviors. *Surf. Coatings Technol.* 377, 124852 (2019). https://doi.org/10.1016/j.surfcoat.2019.08.006.
88. Singh, A., Harimkar, S.P.: Laser surface engineering of magnesium alloys: A review. *JOM.* 64, 716–733 (2012). https://doi.org/10.1007/s11837-012-0340-2.
89. Dubé, D., Fiset, M., Couture, A., Nakatsugawa, I.: Characterization and performance of laser melted AZ91D and AM60B. *Mater. Sci. Eng. A.* 299, 38–45 (2001). https://doi.org/10.1016/S0921-5093(00)01414-3.
90. Mondal, A.K., Kumar, S., Blawert, C., Dahotre, N.B.: Effect of laser surface treatment on corrosion and wear resistance of ACM720 Mg alloy. *Surf. Coatings Technol.* 202, 3187–3198 (2008). https://doi.org/10.1016/j.surfcoat.2007.11.030.
91. Dutta Majumdar, J., Galun, R., Mordike, B.., Manna, I.: Effect of laser surface melting on corrosion and wear resistance of a commercial magnesium alloy. *Mater. Sci. Eng. A.* 361, 119–129 (2003). https://doi.org/10.1016/S0921-5093(03)00519-7.
92. Abbas, G., Liu, Z., Skeldon, P.: Corrosion behaviour of laser-melted magnesium alloys. *Appl. Surf. Sci.* 247, 347–353 (2005). https://doi.org/10.1016/j.apsusc.2005.01.169.
93. Abbas, G., Li, L., Ghazanfar, U., Liu, Z.: Effect of high power diode laser surface melting on wear resistance of magnesium alloys. *Wear.* 260, 175–180 (2006). https://doi.org/10.1016/j.wear.2005.01.036.
94. Qian, M., Li, D., Jin, C.: Microstructure and corrosion characteristics of laser-alloyed magnesium alloy AZ91D with Al–Si powder. *Sci. Technol. Adv. Mater.* 9, 025002 (2008). https://doi.org/10.1088/1468-6996/9/2/025002.
95. Triantafyllidis, D., Li, L., Stott, F.H.: Surface treatment of alumina-based ceramics using combined laser sources. *Appl. Surf. Sci.* 186, 140–144 (2002). https://doi.org/10.1016/S0169-4332(01)00639-0.
96. Mahmood, M., Bănică, A., Ristoscu, C., Becherescu, N., Mihăilescu, I.: Laser coatings via state-of-the-art additive manufacturing: A review. *Coatings.* 11, 296 (2021). https://doi.org/10.3390/coatings11030296.
97. Gutknecht, D., Chevalier, J., Garnier, V., Fantozzi, G.: Key role of processing to avoid low temperature ageing in alumina zirconia composites for orthopaedic application. *J. Eur. Ceram. Soc.* 27, 1547–1552 (2007). https://doi.org/10.1016/j.jeurceramsoc.2006.04.123.
98. Ganesh, I., Olhero, S.M., Torres, P.M.C., Alves, F.J., Ferreira, J.M.F.: Hydrolysis-induced aqueous gelcasting for near-net shape forming of ZTA ceramic composites. *J. Eur. Ceram. Soc.* 29, 1393–1401 (2009). https://doi.org/10.1016/j.jeurceramsoc.2008.08.033.
99. Wang, J., Stevens, R.: Zirconia-toughened alumina (ZTA) ceramics. *J. Mater. Sci.* 24, 3421–3440 (1989). https://doi.org/10.1007/BF02385721.

100. Yan, S., Wu, D., Niu, F., Ma, G., Kang, R.: Al_2O_3-ZrO_2 eutectic ceramic via ultrasonic-assisted laser engineered net shaping. *Ceram. Int.* 43, 15905–15910 (2017). https://doi.org/10.1016/j.ceramint.2017.08.165.

101. Wang, H.M., Yu, Y.L., Li, S.Q.: Microstructure and tribological properties of laser clad CaF_2/Al_2O_3 self-lubrication wear-resistant ceramic matrix composite coatings. *Scr. Mater.* 47, 57–61 (2002). https://doi.org/10.1016/S1359-6462(02)00086-6.

102. Khanna, A.S., Kumari, S., Kanungo, S., Gasser, A.: Hard coatings based on thermal spray and laser cladding. *Int. J. Refract. Met. Hard Mater.* 27, 485–491 (2009). https://doi.org/10.1016/j.ijrmhm.2008.09.017.

103. Li, Y., Bai, P., Wang, Y., Hu, J., Guo, Z.: Effect of TiC content on Ni/TiC composites by direct laser fabrication. *Mater. Des.* 30, 1409–1412 (2009). https://doi.org/10.1016/j.matdes.2008.06.046.

104. Prakash, C., Uddin, M.S.: Surface modification of β-phase Ti implant by hydroaxy-apatite mixed electric discharge machining to enhance the corrosion resistance and in-vitro bioactivity. *Surf. Coatings Technol.* 326, 134–145 (2017). https://doi.org/10.1016/j.surfcoat.2017.07.040.

105. Pradhan, S., Singh, S., Prakash, C., Królczyk, G., Pramanik, A., Pruncu, C.I.: Investigation of machining characteristics of hard-to-machine Ti-6Al-4V-ELI alloy for biomedical applications. *J. Mater. Res. Technol.* 8, 4849–4862 (2019). https://doi.org/10.1016/j.jmrt.2019.08.033.

106. Prakash, C., Kansal, H.K., Pabla, B., Puri, S., Aggarwal, A.: Electric discharge machining – A potential choice for surface modification of metallic implants for orthopedic applications: A review. *Proc. Inst. Mech. Eng. Part B J. Eng. Manuf.* 230, 331–353 (2016). https://doi.org/10.1177/0954405415579113.

107. Prakash, C., Singh, S., Singh, M., Verma, K., Chaudhary, B., Singh, S.: Multi-objective particle swarm optimization of EDM parameters to deposit HA-coating on biodegradable Mg-alloy. *Vacuum.* 158, 180–190 (2018). https://doi.org/10.1016/j.vacuum.2018.09.050.

108. Prakash, C., Kansal, H.K., Pabla, B.S., Puri, S.: Powder mixed electric discharge machining: An innovative surface modification technique to enhance fatigue performance and bioactivity of β-Ti implant for orthopedics application. *J. Comput. Inf. Sci. Eng.* 16, (2016). https://doi.org/10.1115/1.4033901.

109. Prakash, C., Singh, S., Pabla, B.S., Uddin, M.S.: Synthesis, characterization, corrosion and bioactivity investigation of nano-HA coating deposited on biodegradable Mg-Zn-Mn alloy. *Surf. Coatings Technol.* 346, 9–18 (2018). https://doi.org/10.1016/j.surfcoat.2018.04.035.

110. Prakash, C., Singh, S., Pruncu, C., Mishra, V., Królczyk, G., Pimenov, D., Pramanik, A.: Surface modification of Ti-6Al-4V alloy by electrical discharge coating process using partially sintered Ti-Nb electrode. *Materials.* 12, 1006 (2019). https://doi.org/10.3390/ma12071006.

111. Prakash, C., Kansal, H.K., Pabla, B.S., Puri, S.: Multi-objective optimization of powder mixed electric discharge machining parameters for fabrication of biocompatible layer on β-Ti alloy using NSGA-II coupled with Taguchi based response surface methodology. *J. Mech. Sci. Technol.* 30, 4195–4204 (2016). https://doi.org/10.1007/s12206-016-0831-0.

112. Prakash, C., Kansal, H.K., Pabla, B.S., Puri, S.: Processing and characterization of novel biomimetic nanoporous bioceramic surface on β-Ti implant by powder mixed electric discharge machining. *J. Mater. Eng. Perform.* 24, 3622–3633 (2015). https://doi.org/10.1007/s11665-015-1619-6.

113. Prakash, C., Kansal, H.K., Pabla, B.S., Puri, S.: Experimental investigations in powder mixed electric discharge machining of Ti–35Nb–7Ta–5Zrβ-titanium alloy. *Mater. Manuf. Process.* 32, 274–285 (2017). https://doi.org/10.1080/10426914.2016.1198018.

114. Nair, A., Ramkumar, P., Mahadevan, S., Prakash, C., Dixit, S., Murali, G., Vatin, N.I., Epifantsev, K., Kumar, K.: Machine learning for prediction of heat pipe effectiveness. *Energies.* 15, 3276 (2022). https://doi.org/10.3390/en15093276.

115. Tiwari, A.K., Kumar, A., Kumar, N., Prakash, C.: Investigation on micro-residual stress distribution near hole using nanoindentation: Effect of drilling speed. *Meas. Control.* 52, 1252–1263 (2019). https://doi.org/10.1177/0020294019858107.

116. Pramanik, A., Basak, A.K., Littlefair, G., Debnath, S., Prakash, C., Singh, M.A., Marla, D., Singh, R.K.: Methods and variables in electrical discharge machining of titanium alloy – A review. *Heliyon.* 6, e05554 (2020). https://doi.org/10.1016/j.heliyon.2020.e05554.

117. Zheng Yang, K., Pramanik, A., Basak, A.K., Dong, Y., Prakash, C., Shankar, S., Dixit, S., Kumar, K., Ivanovich Vatin, N.: Application of coolants during tool-based machining – A review. *Ain Shams Eng. J.* (2022). https://doi.org/10.1016/j.asej.2022.101830.

118. Usca, Ü.A., Uzun, M., Şap, S., Giasin, K., Pimenov, D.Y., Prakash, C.: Determination of machinability metrics of AISI 5140 steel for gear manufacturing using different cooling/lubrication conditions. *J. Mater. Res. Technol.* (2022). https://doi.org/10.1016/j.jmrt.2022.09.067.

119. Nguyen, D.N., Le Chau, N., Dao, T.P., Prakash, C., Singh, S.: Experimental study on polishing process of cylindrical roller bearings. *Meas. Control.* 52, 1272–1281 (2019). https://doi.org/10.1177/0020294019864395.

120. Biswas, A., Li, L., Maity, T.T., Chatterjee, U.K., Mordike, B.B., Manna, I., Majumdar, J.D.: Laser surface treatment of Ti-6Al-4V for bio-implant application. *Lasers Eng.* 17, 59–73 (2007).

121. Yue, T.., Yu, J.., Mei, Z., Man, H..: Excimer laser surface treatment of Ti–6Al–4V alloy for corrosion resistance enhancement. *Mater. Lett.* 52, 206–212 (2002). https://doi.org/10.1016/S0167-577X(01)00395-0.

122. Li, L., Mirhosseini, N., Michael, A., Liu, Z., Wang, T.: Enhancement of endothelialisation of coronary stents by laser surface engineering. *Lasers Surg. Med.* 45, 608–616 (2013). https://doi.org/10.1002/lsm.22180.

123. Ready, J.F., Farson Dave, F.: *LIA Handbook of Laser Materials Processing.* Laser Institute of America Magnolia Publishing, Orlando, FL (2001).

124. Brandt, M., Sun, S., Alam, N., Bendeich, P., Bishop, A.: Laser cladding repair of turbine blades in power plants: From research to commercialisation. *Int. Heat Treat. Surf. Eng.* 3, 105–114 (2009). https://doi.org/10.1179/174951409X12542264513843.

125. Fan, Y., Wang, Y., Vukelic, S., Yao, Y.L.: Wave-solid interactions in laser-shock-induced deformation processes. *J. Appl. Phys.* 98, 104904 (2005). https://doi.org/10.1063/1.2134882.

126. Fabbro, R., Peyre, P., Berthe, L., Scherpereel, X.: Physics and applications of laser-shock processing. *J. Laser Appl.* 10, 265–279 (1998). https://doi.org/10.2351/1.521861.

127. Sundar, R., Pant, B.K., Kumar, H., Ganesh, P., Nagpure, D.C., Haedoo, P., Kaul, R., Ranganathan, K., Bindra, K.S., Oak, S.M., Kukreja, L.M.: Laser shock peening of steam turbine blade for enhanced service life. *Pramana.* 82, 347–351 (2014). https://doi.org/10.1007/s12043-014-0688-7.

128. Cui, Z., Li, S., Zhou, J., Ma, Z., Zhang, W., Li, Y., Dong, P.: Surface analysis and electrochemical characterization on micro-patterns of biomedical Nitinol after nanosecond laser irradiating. *Surf. Coatings Technol.* 391, 125730 (2020). https://doi.org/10.1016/j.surfcoat.2020.125730.

129. Li, S., Cui, Z., Zhang, W., Li, Y., Li, L., Gong, D.: Biocompatibility of micro/nanostructures nitinol surface via nanosecond laser circularly scanning. *Mater. Lett.* 255, 126591 (2019). https://doi.org/10.1016/j.matlet.2019.126591.

6 Laser Surface Engineering Case Studies

Gaurav Prashar
Rayat Bahra Institute of Engineering and Nano Technology

Hitesh Vasudev
Lovely Professional University

CONTENTS

6.1 INTRODUCTION

Our attitudes and the way we use various technologies are fundamentally changing as society moves toward a "greener" lifestyle. As long as there is no degradation in the quality of the finished products, no one recognizes the advantages of recycling and reusing materials, tools, and technology. Today's engineers must be able to design appealing new goods that utilize all recycling options while also being able to reuse and repair existing appliances in an efficient manner [1]. Fortunately, more mature members of society and the manufacturing sector see the need to cut costs through renovation, regeneration, and repair [2–5]. Industry lacks "free" funds to purchase new equipment; instead, it needs and relies mostly on repairs. Even if the skills were there, no one would be willing to risk their money now [4,6–12]. The mechanical and manufacturing engineering sectors of repair and refurbishment industry are currently expanding as a result of these social shifts, new attitudes, and economic pressure [13,14]. This market for repairs and renovations has vast size and potential.

Laser surface engineering is being applied successfully in the automotive, aviation, shipbuilding and ship repair, military, defense, and many other industries thanks to the rapid expansion of laser applications and the declining cost of laser systems.

The field of laser cladding is a new one in material processing, combining laser technology, CAD/CAM, powder metallurgy, robotics, controllers, and machine design. In next section, few case studies carried out by researchers/professionals in past using laser surface engineering will be discussed.

6.2 CASE STUDY: HYBRID LASER WELDING OF FILLET JOINTS

One of the most difficult problems in welding technology is distortion reduction. When a material is heated and cooled locally during welding, it expands and flows plastically, causing residual stress and distortion. In the shipbuilding industry, where huge components with a relatively low thickness are welded, distortions are particularly severe. Figure 6.1 depicts an illustration of a ship hull with apparent deformities. The vessel's functionality and visual appeal suffer as a result. Through rectification, it also results in an increase in production costs.

One of the most typical forms of joints utilized in shipbuilding are fillet welds. Currently, gas metal arc welding, which applies a consumable electrode in the form of filler wire, is used to create the majority of fillet welds. The size of the weld determines the strength of the joint, although global deformation is related to the heat applied and, consequently, to the amount of deposited material. In comparison with gas metal arc welding, a laser's highly focused energy allows for the energy-efficient welding of a particular thickness. However, the method is susceptible to errors due to challenges in introducing a filler metal and inadequate laser welding fit-up tolerance. Hybrid laser welding (inset Figure 6.2), which combines a laser beam and an arc source, such as gas metal arc welding, offers the best compromise.

In terms of the degree of control over the weld profile, the project conducted by Suder [4] showed the high flexibility of hybrid laser/GMAW welding. For a given thickness, the laser source, or laser power, power density, inclination angle, and welding speed can be used to control the penetration and molten metal distribution.

FIGURE 6.1 Distortion due to welding in a ship hull [15]. (Permissions under Creative Commons Attribution License 3.0.)

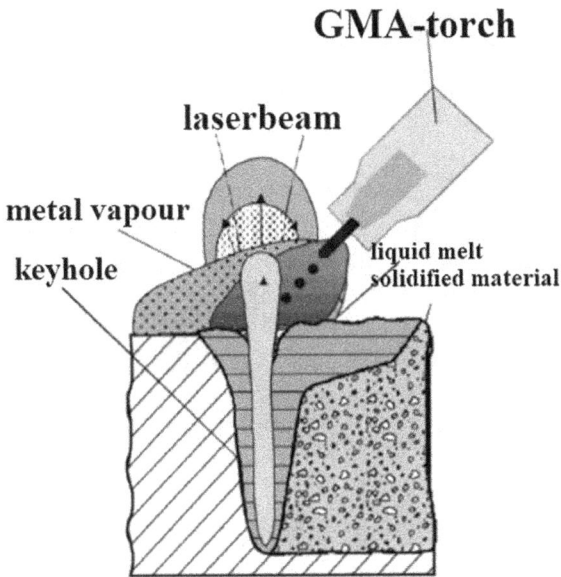

FIGURE 6.2 Schematic of hybrid laser welding process [16]. (Permissions under Creative Commons Attribution License 3.0.)

6.3 REFURBISHING HIGH-VALUE STEAM TURBINE COMPONENTS USING LASER TECHNOLOGY

Reliable and efficient power generation is a major global issue due to both political and environmental concerns. Nevertheless, many critical components, particularly the blades of the low-pressure side of power-generating steam turbines, are subject to failure due to severe erosion at the leading edges. Taking machines offline for maintenance and removal of the damaged blades for repair is extremely expensive. Thereby, increasing the service life of these critical components offers significant economic benefits. Older repair technologies are constantly being challenged by new processes and automation that offer benefits in performance, efficiency, and cost. Repair of turbine components requires cost-effective processes, on-site repair, quick turnaround times, and robust results. The CSIR and Eskom have developed a laser welding process to reconstruct tenons on turbine rotor blade tips (in Figure 6.3), which may lead to significant cost savings at power plants. The objective of the project is to recover lot of un-serviceable blades.

The advantage of laser is low heat input, and this is related to the science of bead, which will be deposited during the process. It is tiny and minute, well build up, can be done accurately with high integrity and cannot be achieved by other conventional techniques. As a result of low heat input and rapid solidification, the mechanical properties achieved were excellent in terms of both strength and impact resistance.

FIGURE 6.3 Laser welding to reconstruct tenons on blade tips.

6.4 ROAD HEADER MINING BOOM RECLAMATION WITH LASER CLADDING

Sustainable demands and cost-effective mining operations were need of the hour where fuel and tooling costs are of major concern. Tooling costs include not only the cost of parts but also production time loss due to unnecessary stops for tool change or the inability to reach a target operation at the appropriate time point. Protecting the ram's hard chrome surface against rock shavings and fine abrasive particles is quite difficult [17–25]. Impacts have the ability to chip and dent the porous hard chromium finish. In damp situations, moisture causes the steel substrate to corrode, which causes the chromium covering to delaminate. The oil seal is harmed by wear and tear to the ram surface, and when these oil seals leak, performance suffers and oil becomes a pollutant. A substantial increase in service life is gained by applying thermal claddings to the wear-exposed regions of ram. The actual cladding material was chosen to accommodate the principal failure mechanism for this application, which is corrosion and impact resistance. Different metallurgy is used to coat internal bearing surfaces. After laser cladding, better corrosion and impact resistance proof surface is obtained, and road headers work for longer durations with little down times and less maintenance costs [26]. This case study was conducted by *Australian-based company named Laser Bond.*

6.5 LASER CLADDING OF TRUCK WHEEL SPINDLES

In Australia's mining industry, big dump trucks (CAT777) are frequently used. Wheel spindles are a high-wear, high-cost component. A growing number of dump trucks are "parked up" in anticipation of maintenance overhauls as a result of the mining downturn. Waiting for new or replacement parts can save a lot of time and money if these and other expensive components are reclaimed using laser cladding. The same laser cladding method was used by ***LaserBond*** to repair these worn-out or damaged surfaces [26]. Because of the metallurgical bond, applied layers can be employed in high-impact, substantially loaded conditions without being at risk of spalling or overlay separation. The laser cladding energy's limitless controllability enables the reduction of unwelcome thermal disintegration of hard phases like carbides, which causes substrate dilution, decomposition, and deformation effects common to alternative restoration techniques.

6.6 TURBINE BLADES REPAIR BY LASER CLADDING

Political and environmental considerations make reliable and efficient electricity generating a crucial global issue. The cost of bringing machines offline for service and removing damaged blades for repair makes it expensive. To overcome this issue, in situ laser cladding of turbine blades was demonstrated worldwide first time by Brandt et al. [27] in 2004. The trial demonstrated that in situ laser cladding of steam turbine blades was possible practically. Shown in Figure 6.4 is the robotic cladding operation at the power station. When the turbine was examined in May 2005, the laser-repaired blades performed well and showed no signs of deterioration. By using more repaired blades and better processing, the October 2005 study significantly boosted confidence in the method. In order to commercialize the invention, Hardwear Pty Ltd was established. In 2007 and 2008, the company successfully completed two commercial contracts.

In the year 2008, Curtis-Wright Corporation., Roseland, reported that Siemens Power Generation utilizes laser shock peening technique to increase fatigue strength of Ti blades of few of their advanced steam turbines [28–34].

FIGURE 6.4   In situ repair of eroded low-pressure blade via laser cladding [27].

6.7 CONCLUSION

These laser-based innovations are anticipated to play a significant role in surface engineering, as there is little question that significantly superior surface qualities are required for modern applications than those in the past. Interest in laser surface engineering is anticipated to persist, and advancements in this field are going on as long as attempts to achieve energy independence and environmental protection are made.

REFERENCES

1. Singh, J., Singh, R., Singh, S., Vasudev, H., Kumar, S.: Reducing scrap due to missed operations and machining defects in 90PS pistons. *Int. J. Interact. Des. Manuf.* (2022). https://doi.org/10.1007/s12008-022-01071-0.
2. Singh, G., Vasudev, H., Bansal, A., Vardhan, S.: Influence of heat treatment on the microstructure and corrosion properties of the Inconel-625 clad deposited by microwave heating. *Surf. Topogr. Metrol. Prop.* 9, 025019 (2021). https://doi.org/10.1088/2051-672X/abfc61.
3. Vasudev, H., Thakur, L., Bansal, A., Singh, H., Zafar, S.: High temperature oxidation and erosion behaviour of HVOF sprayed bi-layer Alloy-718/NiCrAlY coating. *Surf. Coatings Technol.* 362, 366–380 (2019). https://doi.org/10.1016/j.surfcoat.2019.02.012.
4. Prashar, G., Vasudev, H.: A comprehensive review on sustainable cold spray additive manufacturing: State of the art, challenges and future challenges. *J. Clean. Prod.* 310, 127606 (2021). https://doi.org/10.1016/j.jclepro.2021.127606.
5. Prashar, G., Vasudev, H., Bhuddhi, D.: Additive manufacturing: Expanding 3D printing horizon in industry 4.0. *Int. J. Interact. Des. Manuf.* (2022). https://doi.org/10.1007/s12008-022-00956-4.
6. Vasudev, H., Singh, P., Thakur, L., Bansal, A.: Mechanical and microstructural characterization of microwave post processed Alloy-718 coating. *Mater. Res. Express.* 6, 1265f5 (2020). https://doi.org/10.1088/2053-1591/ab66fb.
7. Prashar, G., Vasudev, H.: High temperature erosion behavior of plasma sprayed Al_2O_3 coating on AISI-304 stainless steel. *World J. Eng.* 18, 760–766 (2021). https://doi.org/10.1108/WJE-10-2020-0476.
8. Vasudev, H., Thakur, L., Singh, H., Bansal, A.: A study on processing and hot corrosion behaviour of HVOF sprayed Inconel718-nano Al_2O_3 coatings. *Mater. Today Commun.* 25, 101626 (2020). https://doi.org/10.1016/j.mtcomm.2020.101626.
9. Vasudev, H., Singh, G., Bansal, A., Vardhan, S., Thakur, L.: Microwave heating and its applications in surface engineering: A review. *Mater. Res. Express.* 6, 102001 (2019). https://doi.org/10.1088/2053-1591/ab3674.
10. Prashar, G., Vasudev, H., Thakur, L.: Performance of different coating materials against slurry erosion failure in hydrodynamic turbines: A review. *Eng. Fail. Anal.* 115, 104622 (2020). https://doi.org/10.1016/j.engfailanal.2020.104622.
11. Prashar, G., Vasudev, H.: Structure–property correlation of plasma-sprayed Inconel625-Al_2O_3 bimodal composite coatings for high-temperature oxidation protection. *J. Therm. Spray Technol.* 31, 2385–2408 (2022). https://doi.org/10.1007/s11666-022-01466-1.
12. Vasudev, H., Thakur, L., Singh, H., Bansal, A.: Mechanical and microstructural behaviour of wear resistant coatings on cast iron lathe machine beds and slides. *Met. Mater.* 56, 55–63 (2018). https://doi.org/10.4149/km_2018_1_55.
13. Singh, R., Toseef, M., Kumar, J., Singh, J.: Benefits and challenges in additive manufacturing and its applications. In: Kaushal, S., Singh, I., Singh, S., and Gupta, A. (eds.) *Sustainable Advanced Manufacturing and Materials Processing.* pp. 137–157. CRC Press, Boca Raton (2022). https://doi.org/10.1201/9781003269298-8.

14. Kumar, D., Yadav, R., Singh, J.: Evolution and adoption of microwave claddings in modern engineering applications. In: *Advances in Microwave Processing for Engineering Materials.* pp. 134–153. CRC Press, Boca Raton (2022). https://doi.org/10.1201/9781003248743-8.

15. Soul, F., Hamdy, N.: *Numerical Simulation of Residual Stress and Strain Behavior After Temperature Modification.* InTech (2012). DOI: 10.5772/47745

16. Zhou, J., Tsai, H.L.: Hrbrid laser-arc welding. In: *Welding Processes.* InTech (2012). https://doi.org/10.5772/50113.

17. Singh, J., Gill, H.S., Vasudev, H.: Computational fluid dynamics analysis on effect of particulate properties on erosive degradation of pipe bends. *Int. J. Interact. Des. Manuf.* (2022). https://doi.org/10.1007/s12008-022-01094-7.

18. Singh, J., Kumar, S., Mohapatra, S.K.: Study on solid particle erosion of pump materials by fly ash slurry using taguchi's orthogonal array. *Tribol. - Finnish J. Tribol.* 38, 31–38 (2021). https://doi.org/10.30678/fjt.97530.

19. Singh, J.: Application of thermal spray coatings for protection against erosion, abrasion, and corrosion in hydropower plants and offshore industry. In: Thakur, L. and Vasudev, H. (eds.) *Thermal Spray Coatings.* pp. 243–283. CRC Press, Boca Raton (2021). https://doi.org/10.1201/9781003213185-10.

20. Singh, J., Singh, S., Pal Singh, J.: Investigation on wall thickness reduction of hydropower pipeline underwent to erosion-corrosion process. *Eng. Fail. Anal.* 127, 105504 (2021). https://doi.org/10.1016/j.engfailanal.2021.105504.

21. Singh, J., Mohapatra, S.K., Kumar, S.: Performance analysis of pump materials employed in bottom ash slurry erosion conditions. *J. Tribol.* 30, 73–89 (2021).

22. Singh, J.: A review on mechanisms and testing of wear in slurry pumps, pipeline circuits and hydraulic turbines. *J. Tribol.* 143, 1–83 (2021). https://doi.org/10.1115/1.4050977.

23. Singh, J., Singh, S.: Neural network prediction of slurry erosion of heavy-duty pump impeller/casing materials 18Cr-8Ni, 16Cr-10Ni-2Mo, super duplex 24Cr-6Ni-3Mo-N, and grey cast iron. *Wear.* 476, 203741 (2021). https://doi.org/10.1016/j.wear.2021.203741.

24. Singh, J., Singh, J.P.: Numerical analysis on solid particle erosion in elbow of a slurry conveying circuit. *J. Pipeline Syst. Eng. Pract.* 12, 04020070 (2021). https://doi.org/10.1061/(asce)ps.1949-1204.0000518.

25. Singh, J.: *Investigation on Slurry Erosion of Different Pumping Materials and Coatings.* Thapar Institute of Engineering and Technology, Patiala, India (2019).

26. LaserBond Limited. *Mining Boom Cylinder Reclad & Repair.* (2021). https://laserbond.com.au/case-studies/mining/item/mining-boom-cylinder-reclad-repair.html

27. Brandt, M., Sun, S., Alam, N., Bendeich, P., Bishop, A.: Laser cladding repair of turbine blades in power plants: From research to commercialisation. *Int. Heat Treat. Surf. Eng.* 3, 105–114 (2009). https://doi.org/10.1179/174951409X12542264513843.

28. Sundar, R., Pant, B.K., Kumar, H., Ganesh, P., Nagpure, D.C., Haedoo, P., Kaul, R., Ranganathan, K., Bindra, K.S., Oak, S.M., Kukreja, L.M.: Laser shock peening of steam turbine blade for enhanced service life. *Pramana.* 82, 347–351 (2014). https://doi.org/10.1007/s12043-014-0688-7.

29. Prakash, C., Kansal, H.K., Pabla, B.S., Puri, S.: Multi-objective optimization of powder mixed electric discharge machining parameters for fabrication of biocompatible layer on β-Ti alloy using NSGA-II coupled with Taguchi based response surface methodology. *J. Mech. Sci. Technol.* 30, 4195–4204 (2016). https://doi.org/10.1007/s12206-016-0831-0.

30. Pramanik, A., Basak, A.K., Littlefair, G., Debnath, S., Prakash, C., Singh, M.A., Marla, D., Singh, R.K.: Methods and variables in Electrical discharge machining of titanium alloy – A review. *Heliyon.* 6, e05554 (2020). https://doi.org/10.1016/j.heliyon.2020.e05554.

31. Pradhan, S., Singh, S., Prakash, C., Królczyk, G., Pramanik, A., Pruncu, C.I.: Investigation of machining characteristics of hard-to-machine Ti-6Al-4V-ELI alloy for biomedical applications. *J. Mater. Res. Technol.* 8, 4849–4862 (2019). https://doi.org/10.1016/j.jmrt.2019.08.033.

32. Prakash, C., Kansal, H.K., Pabla, B.S., Puri, S.: Experimental investigations in powder mixed electric discharge machining of Ti–35Nb–7Ta–5Zrβ-titanium alloy. *Mater. Manuf. Process.* 32, 274–285 (2017). https://doi.org/10.1080/10426914.2016.1198018.

33. Prakash, C., Uddin, M.S.: Surface modification of β-phase Ti implant by hydroaxyapatite mixed electric discharge machining to enhance the corrosion resistance and in-vitro bioactivity. *Surf. Coatings Technol.* 326, 134–145 (2017). https://doi.org/10.1016/j.surfcoat.2017.07.040.

34. Prakash, C., Singh, S., Pruncu, C., Mishra, V., Królczyk, G., Pimenov, D., Pramanik, A.: Surface modification of Ti-6Al-4V alloy by electrical discharge coating process using partially sintered Ti-Nb electrode. *Materials.* 12, 1006 (2019). https://doi.org/10.3390/ma12071006.

7 Applications of Laser Technology for Processing and Post-Treatments of Powders and Coatings

Jashanpreet Singh
Chandigarh University

CONTENTS

7.1 LASER TECHNOLOGY FOR COATINGS AND CLADDINGS: INTRODUCTION

The issue of safeguarding machines against various forms of damage or failure is a current challenge in almost all branches of engineering [1–6]. Many engineers, researchers, and scientists acknowledge the need of safeguarding a wide range of industrial machinery, equipment, and systems [7]. Economical solutions to such

DOI: 10.1201/9781003347408-7

failure situations need a deep knowledge of the underlying processes involved [8]. There has been significant expansion in the materials industry recently. It is possible to find materials that can hold up under a wide range of stresses, including tensile, compressive, impact, fatigue, and so on [8–11]. The mechanisms of deterioration, however, are proven to be more destructive to materials. For this reason, balancing performance, maintenance, and cost becomes a herculean challenge in material selection [7,12–17]. Adverse effect of material degradation processes is predominant in industrial as well social sector. Most common degradation processes are fatigue failure, erosion, corrosion, and abrasion [3,12,15,16,18–40]. Therefore, the surface protection techniques are most feasible to protect the surface of machinery against the various degradation processes [25,39,41–43]. Most commonly used surface protection techniques are coatings and claddings [44–54]. Coatings can be deposited by using different methods such as thermal spray [55–62], cold spray [2], arc spray [8], plasma spray [63–65], D-gun spray [66], chemical deposition [67–69], electroplating [21,70], and claddings [71].

The Institute of Theoretical and Applied Mechanics of the Russian Federation is credited with pioneering the technology that made cold spray additive manufacturing (AM) possible in the mid-1980s [72]. Before the 1980s, thermal spray technologies were considered "tough to manage," but since then, they have been tightly governed. There was an initial emphasis on developing thermal spray-based coatings for use in the hard tool sector. Powder, wire, or rod is fed into a spray gun's flame to create a coating, and the process is often described as one in which the coating is generated from melted or partially melted droplets [73]. Coatings or components for 3D AM may be made with qualities that are unique to themselves due to the nature of the energy source (whether kinetic or thermal) and the nature of the material that is deposited [2].

Coatings, in particular, are used in the repair and production of components across the board in the engineering industry [15,22]. To add, coatings are used to enhance the surface attributes of components at the choice of a low-cost, lightweight, and sustainable investment. HVOF (high-velocity oxy fuel) and plasma spraying are two examples of evolution that have proven especially beneficial in producing coatings with lower porosity and good adherence [74]. These novel coatings are used to give features that are superior to those of untreated substrates in order to attain a better degree of performance in a particular application, or to provide an anticipated increase of the in-service lifetime of a component [75,76]. Numerous properties, such as fatigue strength, clearing control, self-healing, self-cleaning, non-stick, water repellency, antimicrobial, biocompatible, bioactive, diffusion barrier, and high absorption capacity [77,78], need the use of advanced coatings on components. Gas turbine engines, upstream pumps, drill bits, architectural glass, orthopaedic implants, and many more products rely on disruptive technologies that pave the way for future developments.

LPBF (laser powder bed fusion), EBM (electron beam machining), and DED (directed energy deposition) are all powder bed-based technologies often utilised to create dense metallic components [79,80]. This rapid melting and solidification of the powder are achieved by the interaction of the supplied powder with the laser or electron beam that produces the melt pool [81]. High-temperature gradients and sizable cooling rates are seen [82,83] because the time spent in contact is short and the heat input is concentrated in a small area. This is because of two factors. These parameters have an effect on the as-built microstructures as well, which results in significant levels of residual stress [84,85]. As a result of the unavoidable faults, the mechanical and fatigue characteristics of the component will continue to deteriorate [86–88]. LPBF has come a long way since its start in the 1990s, thanks in large part to the needs and expectations of many industries including aerospace, biomedicine, the military, and the automotive industry [79,89]. This fashion will continue to gain momentum in the years to come, providing manufacturers with a helping hand as they make the move toward production that is more inventive, digital, and high value. This section discusses both the present status of the LPBF technology and its potential in future. This chapter focuses on the most recent developments in LPBF technology and its implications. Recent developments in laser technology, novel metal powders and alloys, and post-processing enhancements are the few examples of these topics [90].

7.2 LASER POWDER BED FUSION (LPBF)

LPBF is an AM technique that has revolutionised the manufacturing business. This is because it enables for the cost, time, and labour involved in the production of complicated components to be reduced. It is possible to manufacture complicated forms with this technology without the need for tooling, castings, or other traditional production techniques. The LPBF process is very necessary for comprehending the connection that exists between the operating parameters and the qualities of the finished product. One of the ways for fusing powder beds is called "low-pressure bed fusion" (LPBF). During this procedure, a bed of powder is first extended out, and then targeted areas are exposed to intense laser light. It allows particles to be melted and fused layer by layer in line with a CAD-generated design. The very name of the procedure conveys a great deal of information about it. The employment of laser energy as the heat source is indicated by the term "laser," powder melting by the word "melting," and selective heating by the phrase "selective" [91]. A laser source, platform, automated system to distribute powder, a controlling system, and rollers are often included in the LPBF system configuration [92]. A beam deflection system that consists of mirrors and focusing lens is responsible for monitoring both the movement of the high-intensity laser beam and its focus.

Figure 7.1 illustrates the many parts that make up an LPBF machine. In LPBF, the powder particulate is completely melted by applied heat in the form of the laser beam, which is analogous to the concept behind the welding process. As a result, this method belongs to the category of deposition welding processes [93,94]. When light is converted into heat, a melt pool forms, and due to surface tension, the pool eventually assumes the shape of a cylinder with slits or a sphere [95]. Due to the relatively brief interaction time period between the powder bed and laser, transient temperature fields with temperatures as high as $105°C$ are produced. Subsequently, a fast quenching effect also takes place with cooling rates as high as $106–108°C/s$ [96]. The effects of the quick solidification process include metallurgical phenomena such as refining microstructures, solid-solution hardening, and metastable phase development, all of

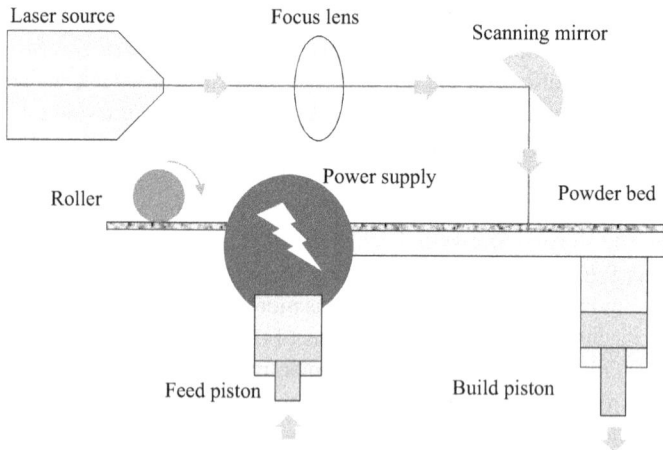

FIGURE 7.1   Schematic of a typical LPBF machine.

those have the potential in improving the mechanical properties of newly generated parts [97].

The fabrication of completely dense components is the primary purpose of the LPBF approach. Because the LPBF process does not include any kind of mechanical pressure and the fluid dynamics are likewise solely controlled by gravity and the capillary forces, achieving such a result might be challenging. This is due to the fact that there is no mechanical pressure involved in the process. Without mechanical pressure, components solidify more easily. This causes track melting and poor, uneven surfaces [98]. High-temperature fluctuations during LPBF enhance residual stress in fast-solidifying layers [99]. A highly rapid cycle of heating and cooling causes the development of heat-affected zones (HAZs) in the area around the melt pool. This HAZ has the potential to change the microstructure as well as the content of the material, which is ultimately what determines the quality of the finished product [91]. Processing parameters have the ability to exert control on the thermal behaviour at any given point in time. The spacing of the hatches, the thickness of the layers, the scanning speed, the laser power, and the scanning technique are the processing parameters. Many parameters of the LPBF process are shown in Figure 7.2. The settings that were selected make it possible for the powder to completely melt and fuse with the layer that came before it. Inappropriate selection of these factors might result in undesired consequences such as thermal cracking, balling, or porosity, or it could signal other undesirable impacts. Developing a relationship between these elements and output outcomes is required to optimise processing settings and achieve the desired result.

7.3 MATERIALS USED IN LPBF

This section will elaborate the current efforts and developments made in the field of processing functional materials using the AM technique in order to build industrial goods.

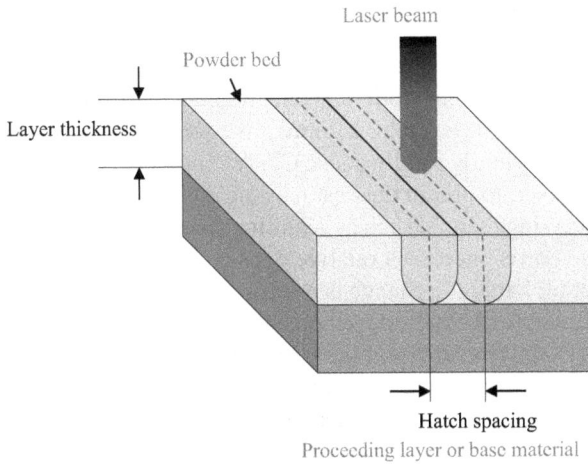

FIGURE 7.2 Schematic of LPBF processing parameters.

7.3.1 TITANIUM ALLOYS

Titanium (Ti) alloys have a high strength-to-weight ratio, are compatible with other materials, and are resistant to high temperatures [100,101]. This has led to their widespread use in a variety of industries, including the biomedical, automotive, and aerospace sectors. Materials utilised in biomedical fields are often required to meet the requirements of lower Young's modulus, good strength, low density, higher erosion resistance, good corrosion resistance, and excellent compatibility [1,21,70,102]. These alloys are an ideal option for use in the biomedical sector [103–106] due to their outstanding qualities, which are stated above. Dental fields, joint replacements, implants, orthodontic components, surgical equipment, and artificial heart valves are only some of the biomedical applications that may make use of these materials. Ti alloys have historically been difficult to manufacture due to the extensive amounts of time, energy, and material that are required. However, the development of AM, in particular LPBF, gives rise to an increase in the production of titanium components [107–110]. Ti made from commercially pure (CP) titanium has been phased out as an orthopaedic prosthesis in favour of Ti6Al4V due to the latter material's superior mechanical properties [111,112]. Because of their lack of toxicity, other titanium alloys are sometimes used in place of CP-Ti and Ti6Al4V [113–115]. These other titanium alloys include niobium (Nb) and zirconium (Zr), which are the two elements that are alloyed together to create Ti-13Nbe-13Zr, which is proving to be an outstanding alloy with exceptional qualities. This alloy is a convenient alternative to Ti6Al4V. In the presence of Nb, the development of the alpha phase is inhibited [116–118]. Strong arguments for using Ti13Nb13Zr instead of Ti6Al4V have been presented in a large number of scholarly articles [119–127]. LPBF lattice structures have also been investigated by a number of researchers with regard to their mechanical performances [128–130] and their porous biomaterials [130–132].

7.3.2 A-TI-BASED MATERIAL

The degree to which the process parameters are optimised has a significant influence on the mechanical behaviour, densification, and morphology of the finished item. An energy density (E) of 1.2 kJ/mm^3 has been proven as being adequate for the fabrication of virtually totally dense CP-Ti components. This was determined via testing. However, to obtain the best feasible density, adequate optimisation of additional processing factors, such as scanning speed and laser power is required. Increased laser power increases relative density when power density stays constant. On the other hand, the change is not consistent everywhere. The processing parameters ultimately affect the development of the microstructure during the process of LPBF [133]. Depending on the parameters used during fabrication, CP-Ti may exhibit microstructural behaviour anywhere from plate-like α-martensitic to acicular martensitic α'-phase. Microstructure may be distinguished in large part by increasing the scanning speed, as has been seen. Transition from the α-phase to the β-phase happens entirely during solidification as a consequence of energy thermalisation inside the melt pool when the scanning speed is less than 10 cm/s and the energy density is larger than 1.2 kJ/mm^3. In contrast, the formation of an α' microstructure takes place when the speed of scanning is greater than 10 cm/s, despite the fact that the energy density remains the same. This is because of the rise in thermal as well as kinetic undercooling. The α'-martensite in columnar prior β-grains produced by Ti6Al4V is an LPBF microstructure that has been observed by a lot of investigators [92,128,129,131,132,134]. The creation of such microstructures is a consequence of the processing settings used. As a consequence, α'-martensite may form if the material is cooled at a rate of more than 410 K/s below the temperature at which martensite formation first started [135]. The build direction of β-grain will become more elongated as a result of heat conduction [134]. It has been proven that LPBF-manufactured CP-Ti components have improved mechanical characteristics compared to their conventionally manufactured counterparts [136–138]. It is possible that the grain refinement that occurs throughout the LPBF results in superior mechanical properties.

7.3.3 β-TI ALLOYS

Ti24Nb4Zr8Sn, more generally referred to as "Ti2448," is regarded a typical β-Ti alloy since it has exceptional qualities such as higher strength and lower modulus. In addition, the processing parameters for this alloy need to be tuned in order to produce completely dense pieces [139–144]. An increase in scanning speed reveals a progressive reduction in both the hardness and density of the object. The conclusion that can be drawn from this is that the processing parameters have a significant impact on the microhardness and density of the material. Figure 7.3 displays two distinct groups of densities for LPBF-manufactured parts: nearly dense (99.3%) and transitional dense (98.2%). These densities were achieved by operating the scanner at two distinct speeds. Figure 7.3 has nothing but laser tracks, which may be seen as the black bands throughout it. Parts made by rolling and forging have somewhat higher

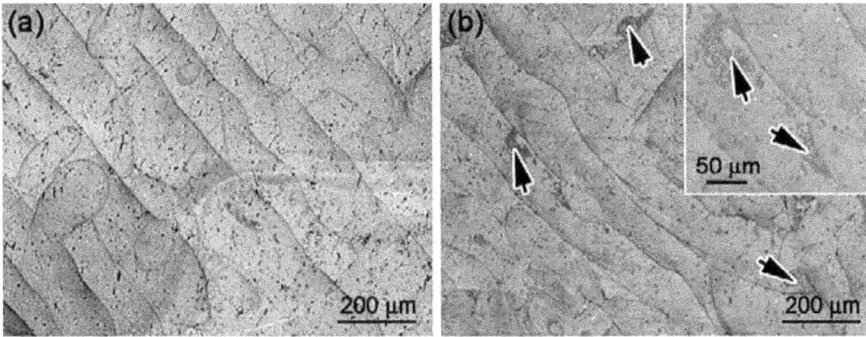

FIGURE 7.3 SEM image of Ti2448 fabricated using LPBF process having relative density: (a) 99.3% and (b) 98.2% [139]. (Permissions from ©Elsevier.)

ultimate and yield strengths than LPBF-made ones, according to the specifications of Ti2448 components made using both techniques [139].

7.3.4 (α + β)-Ti Alloys

By using a microstructure known as "lamellar (α+β)," Ti6Al4V's ductility may be increased; however, this does not have to come at the price of the material's yield strength [145]. Altering the energy density and making use of the cyclic reheating that is associated with layer deposition are both required in order to change lamellar (α + β) into a' martensite [146]. The majority of the LPBF research on (α, β) Ti has been conducted on Ti6Al4V [147,148] and Ti6Al7Nb [149,150]. SEM microstructure analysis of Ti6Al4V components manufactured using LPBF shows fine acicular martensite CP-Ti due to the large temperature gradients present during the process. The process parameters are required to be optimised in order to generate the most dense parts feasible. The components produced using LPBF have better attributes to those created using other production processes. During the LPBF process, martensite microstructure is thought to have been formed, which is responsible for this result [147]. Ti6Al7Nb is an additional Ti-based alloy that is utilised for biomedical implants. In comparison with Ti6Al4V, it has advantageous characteristics such as increased resistance to corrosion and improved mechanical qualities [151,152]. Ti6Al7Nb and Ti6Al4V that are manufactured by LPBF have a martensite microstructure that is quite close to one another [153–158]. Ti6Al7Nb components manufactured by LPBF have improved characteristics compared to those obtained via casting. Due to their wide range of high-value applications, aluminium alloys have become one of the most studied material systems in selective laser melting (SLM). However, processing these alloys is difficult because of the challenges associated with laser-melting aluminium, which results in components with a variety of defects. In this decade, numerous researchers developed different methods to solve these issues, reporting successful SLM of a variety of aluminium alloys and studying its possible use in more complex componentry [159].

7.3.5 COMPOSITE AND POROUS TI MATERIAL

Any study that has been done on LPBF and Ti alloys has mostly been limited to well-known alloys. On the other hand, ceramics such as SiC, TiC, or TiB_2 are added to titanium in order to boost its wear resistance, yield strength, and ultimate strength [160,161]. As shown in Figure 7.4, the incorporation of titanium monobromide (TiB) as reinforcement contributes to the material's chemical, thermodynamic, and mechanical stability. The production of Ti-TiB composite occurs as a result of an laboratory condition reaction involving Ti and titanium dibromide (TiB_2) [162,163]. Adjusting the process conditions and using a Ti-5wt% TiB2 combination powder allowed Attar et al. [164] to create the densest Ti-TiB component. Needless morphology was discovered to be ubiquitous in the Ti matrix. Ti-increased TiB's microhardness may be related to the hardening effect that is brought about by the refining of the Ti grains, as shown by a comparison of features of LPBF-fabricated titanium composites. And the reason why Ti-TiB has such a high yield and final strength is because of the strengthening impact that the TiB particles have.

In the field of biomedicine, the fundamental objective of using porous titanium material is to simulate the properties of genuine bone. While LPBF is being done on porous Ti alloys, the CP-Ti and Ti6Al4V are the primary centres of focus that are being paid. Compressive testing was performed on a 55%–75% porous titanium structure that was manufactured via LPBF [165]. This structure is comparable to the natural human bone. Attar et al. [166] were able to effectively synthesise porous Ti-TiB and CP-Ti using LPBF. The researchers achieved porosity values of 10%,

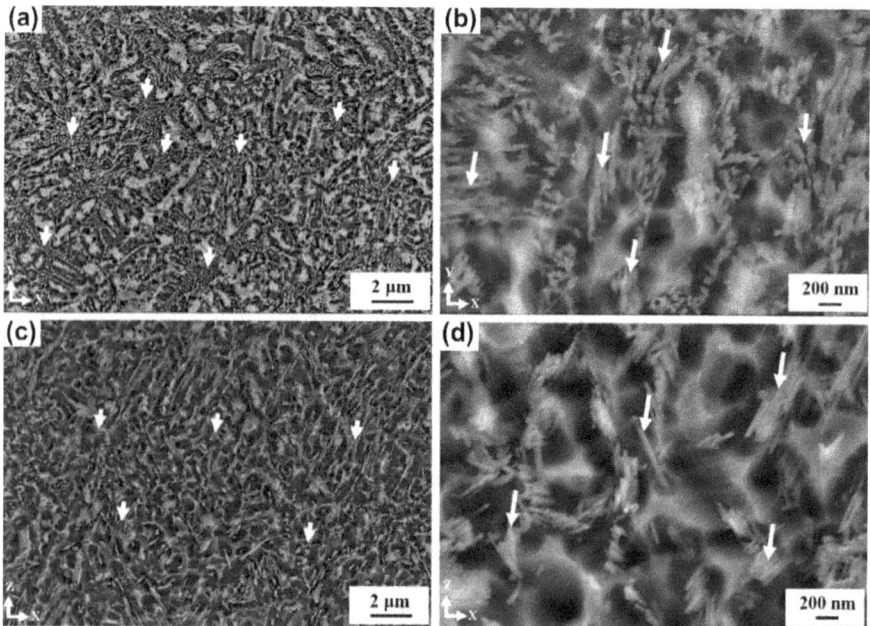

FIGURE 7.4   SEM of the SLM-based Ti-TiB composite along (a, b) cross-sectional and (c, d) longitudinal axis at different magnifications [160]. (Permissions from ©Elsevier.)

17%, and 37%, respectively. Because Young's modulus of these materials and their yield strength were somewhat comparable to those of human bones, it is possible that they may be used as an alternative to implants. The fundamental objective of the LPBF is to create a strong, biocompatible, porous Ti-based implant that promotes osseointegration in the patient's body [167].

7.3.6 MAGNESIUM ALLOYS

Magnesium (Mg) ranks as the sixth most readily accessible element in the crust of the planet. Because of their low structural weight, Mg alloys are favoured for weight-sensitive applications [168]. Mg alloys are lighter than other Al or Ti. Mg is attracting interest from a variety of industries, including the aviation and automotive sectors [169,170]. This is due to the fact that the element is both lightweight and high strength. In addition, the alloy based on magnesium has outstanding mechanical qualities, including castability and machinability, high thermal stability, and high thermal and electrical conductivity [170–173]. The difficulty with magnesium, however, is that it has a low resistance to corrosion, in addition to other undesirable qualities such as a low Young's modulus, lesser strength, and lower creep resistance [174,175]. Because of its low formability, fast degradation, and hydrogen evolution [176], magnesium is still rarely used to its full potential in the therapeutic sector. As a result, scientists are always developing new magnesium alloys and composites in order to satisfy the needs of various jobs. Because it has both high strength and resistance to corrosion, Mg alloy, namely Mg-Zn, is considered to be the superior alloy [177–179]. The permanent mould casting [180], extrusion [181], rolling [182], and pressing [183] processes are the ones that create the majority of the Mg-Zn products. However, since the Mg matrix has a closed-packed hexagonal structure (HCP), its formability is quite poor [184]. Because of this, the plastic working of these alloys has to be done at higher temperatures than the surrounding area, which drives up the cost of production. The quick manufacture of such alloys as high-density goods has been made possible by the development of AM, particularly LPBF, which eliminates the need for any moulds or fixtures in the manufacturing process [79].

7.3.7 AL-BASED MATERIALS

Next to steel and iron, aluminium (Al) and Al alloys are the most widely utilised metals in the construction industry. Applications in aircraft, power electronics, vehicles, and weapons development all benefit from their exceptional features of high strength, low density, and outstanding corrosion resistance [185–187]. After steel and iron, Al and its alloys are the most used building materials. In the production of any object made from aluminium or an alloy of aluminium, the traditional manufacturing processes of extrusion, casting, and forging are often used [188]. Despite the fact that these techniques have been shown to be excellent ways for the production of aluminium, there are still a great deal of downsides connected with them. When casting, coarse microstructures are created because of slow cooling rates and a high number of flaws [189,190]. Porosity, slag inclusion, and offset flaws are only a few examples. The manufacturing process loses some of its adaptability due to the isolation of the high-performance aluminium alloys assembly line. As a result of the progress made in today's sectors, the standards for the performance of aluminium components are among the most stringent. As a result, structures are necessary in order to satisfy such criteria and must be produced in order to cut down on both cost and time [191]. It is generally knowledge that LPBF produces satisfactory results of completely dense sections in a single scan on its own. However, in the case of specific alloys like Al/Fe_2O_3 powders, in situ development of particle-reinforced Al matrix was required to overcome defects like balling, doss formation, and part deformation, all of which result in fabrication of part with poor surface finish. This was the case because these defects caused the part to have a poor surface finish. In addition, process parameters play a vital role in determining the morphology, fracture behaviour, and cyclic fatigue of alloys that are manufactured using LPBF, such as AlSi10Mg [192]. This is the case regardless of whether the alloy is wrought or cast. In the not too distant future, the manufacturing sector may have a significant amount of interest in the techniques of production for aluminium alloys that have different designs, high precision, and structures that are close to net form. Many of the issues that were associated with the conventional methods of production have been resolved as a result of the application of the LPBF process to aluminium [193–195]. According to Louvis et al. [196], the combination of high laser power and slow scanning speed is what leads to the formation of unstable huge melt pools, which in turn leads to balling. This not only lengthens the manufacturing time but also drives up the costs. The balling problem may be remedied by using low laser energy and high scanning speeds in the manufacturing process. It is not feasible to get rid of the influence that oxides have on LPBF for aluminium. As a result, there is a need for more research effort to be done in order to develop ways to regulate the development of oxides. Commercially accessible Al alloys are used in manufacturing, in terms of applications, essential features, and heat treatment, arrangement of compositional elements in series [157].

7.3.8 Hard Materials

The production of hard metals is challenging because the extensive amounts of time, energy, and materials are required. The development of SLM has made the manufacture of these materials far more convenient. When trying to draw any kind of conclusion regarding the microstructures or mechanical characteristics of the fabricated components, it is crucial to have a firm grasp on the relationship that exists between the increased density of the manufactured part and the process parameters. Because of the qualities that they possess, titanium, magnesium, and aluminium are among the metals that are given the most consideration for usage in a variety of sectors, including aerospace and biomedical. It is possible that the differences in the laser parameters, particularly the laser scanning speed, are to blame for the dissimilarities in the microstructure of Ti. In the instance of magnesium, the microstructural development is tracked using the input and cooling rates of a particular kind of laser energy. LPBF may create samples with better qualities to those made using conventional procedures like casting, as shown by the microhardness, compressive strength, hardness, and tensile strength of LPBF-fabricated samples. These mechanical characteristics include things like tensile and compressive strength, as well as hardness and microhardness. It has also been discovered that the incorporation of additional metals into an alloy during the LPBF processing phase results in the production of a product that is superior in terms of efficiency.

7.4 PROCESSING OF POWDERED MATERIALS

7.4.1 Processing of Mg-Based Materials

There is a possibility that magnesium granules might be used in the LPBF process while Mg rails are melted entirely in an inert environment [197]. Single-track Mg powders' processing window was determined by observing the powder's response to the laser's energy under varying conditions [198]. This interaction was studied to determine how different sets of processing parameters would affect the processing window. Numerous researches have also looked at the formability of magnesium and its alloys such as AZ91D, WE43, ZK60, and Mg-9%Al to create single-layer and multi-layer 3D structures, with the goal of pinpointing the processing window for these materials [199]. The processing map for the Mg alloy with 9% Al is shown in Figure 7.5. The laser energy density is used for purpose of influencing the occurrence of the regions and microstructures as well as comparing the processing conditions for single- and multiple-layer components. When looking at the behaviour of magnesium powder and its alloys throughout a range of scanning speeds and laser powers, there are four distinct zones that can be identified based on the quality of the components that are manufactured using LPBF. You may find an explanation for them below.

The energy density and laser power in this area are both much higher than required scanning speed. Due to the fact that magnesium has a low melting point, the

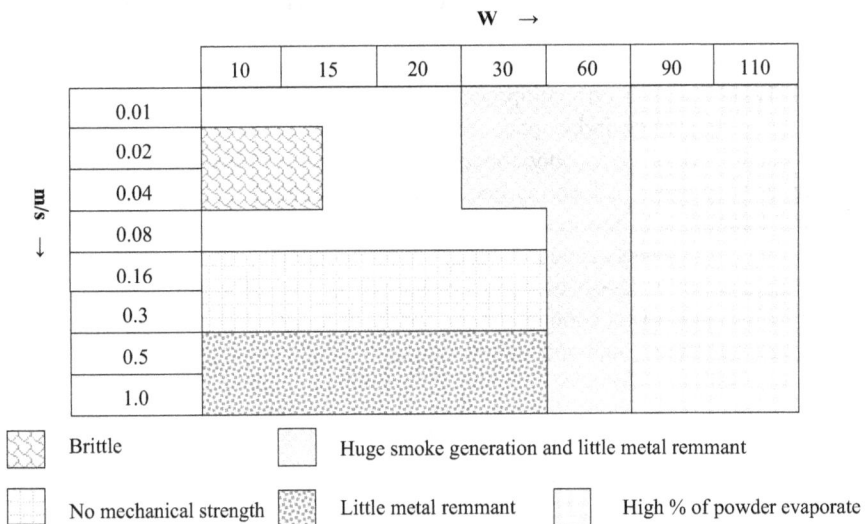

FIGURE 7.5   Processing of Mg-9%Al with respect to laser power and scanning speed.

powdered magnesium that is present in the melt pool is subject to evaporation and ionisation as a consequence of the high temperature. The vaporised particles cause an expansion that results in a tremendous recoil effect inside the melt pool. This impact causes the liquid and the powder to be blown away, which ultimately leads to the development of no track [200]. The melt pool becomes unstable because the high temperature also changes the viscosity of the heated liquid Mg. The enormous amount of energy input also results in substantial thermal strains, which ultimately causes the pieces to become deformed [201].

Both the energy density and laser power here are well below what would be needed for any practical scanning speeds. Low energy input and rapid scanning would cause only partial melting of Mg particles because of insufficient contact time. Insufficient liquid-phase formation due to insufficient laser energy leads to a weak neck of attraction between particles. This is because the liquid phase requires a higher energy level. In situations like these, fusion between the particles results in portions that have little mechanical strength, and a large number of pieces of debris that have not melted may be detected on the surface [201]. When heat is carried out from the centre of the melt pool and into the powders that surround it, a heat-affected zone, also known as a "HAZ," is created as a result of partial melting [198]. MgO_2, popularly known as "black fog," would form if the scanning speed was raised further due to the low density and chemical characteristic of the magnesium particles, which may disturb the processing conditions [202].

Mg powders may be allowed to melt to a significant degree inside this region, with the resulting melt pool being much more stable and producing tracks with strong bonding between the particles. Powder bed heating and melt pool viscosity reduction are both greatly facilitated by the laser energy spectrum present in this region. As a result, the melt is able to spread evenly over the pre-processed layer, which results in a denser component. When the laser power is lower and the scanning speed is faster, the particles' surfaces melt due to the massive quantity of low energy input. This results in a weak binding neck between the particles. As can be seen in zone [199], the created samples exhibited a powder stacking structure under the circumstances of the laser, but they lacked any kind of mechanical strength. Experiments conducted with CP-Ti [149] and Ti-TiB$_2$ components shown that LPBF in this area is capable of producing parts with desirable qualities [164].

The occurrence of the balling area is what distinguishes this one from the others. Balling is the process of forming enormous melt pools by agglomerating powders that have been melted into the shape of balls. This is mostly the consequence of a lower laser energy density input, which amounts to little more than a high layer thickness, a rapid scanning speed, and a low power output [100]. It is possible that the surface of the manufactured component may suffer damage if the balling phenomenon is allowed to continue occurring in this zone.

7.4.2 Processing of Al-Based Materials

Laser power was varied from 20 to 240 W, scanning speed was varied from 20 to 250 mm/s, and the hatch spacing was held constant at 0.1 mm throughout all experiments [203,204]. The results of these experiments were used to create a map of processing parameters for Al and Al alloys, with a particular focus on pure aluminium,

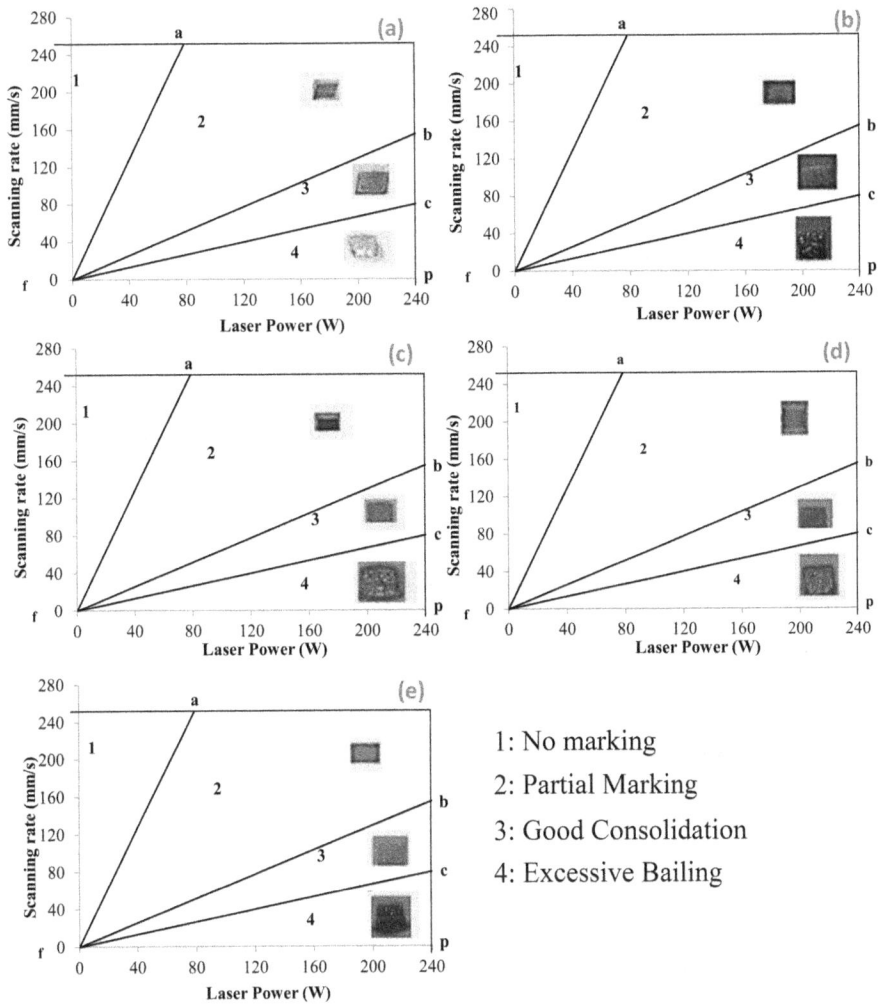

FIGURE 7.6 Processing map for monolayer components generated in (a) 100% Al atomised with air, (b) 100% Al atomised with gas, (c) Al-5.6 Mg atomised with H_2O, (d) Al-6Mg atomised with H_2O, and (E) Al-12Si atomised with gas [203]. (Permissions from ©Elsevier.)

Al-Mg, and AlSi$_{12}$. It was determined that there were four regions: those with no marking, those with incomplete marking, those with excellent consolidation, and those with severe balling (Figure 7.6). A consistent pattern can be seen across all of the powders in representation of the data in Figure 7.7. The borders that separated the various areas had certain subtle nuances that made them distinct from one another. The areas of incomplete marking mostly consisted of components that were manufactured with an exceptionally low level of strength. The area that had strong consolidation produced samples that had high strength and good coherent bonding. These

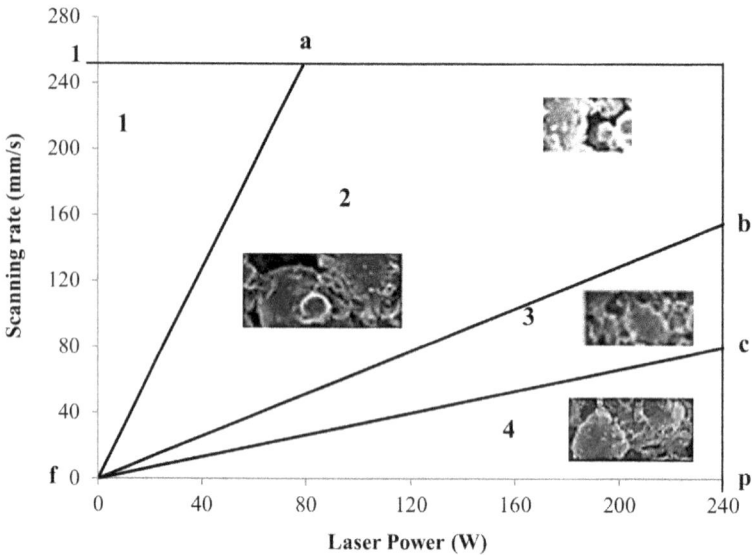

1: No marking; 2: Partial Marking; 3: Good Consolidation; 4: Excessive Bailing

FIGURE 7.7 Processing of Al powder display various microstructures at different scanning rate and laser power [203]. (Permissions from ©Elsevier.)

characteristics were achieved throughout the manufacturing process. As a result, an area with strong consolidation needs to be deemed the right location for producing multi-layered components using LPBF. The area that has an excessive amount of balling also has an undesirable fault called "balling," which causes enormous melt pools. The next paragraphs will examine the factors that led to the formation of such areas.

The area of the material where there is no marking is affected by the low energy density as well as the brief contact time period that occurs between the laser and the material. This may result in weak interparticle bonding. During the LPBF process, poor bonding may also be the consequence of employing fast scanning rates in conjunction with low laser powers.

The zone of incomplete marking is distinguished by the presence of produced pieces that have a significant number of porous faults. This is also a result of the implementation of low energy density, which is a consequence of the use of fast scan speeds and low laser powers. Because of the way these processing parameters are combined, an insufficient liquid phase is formed, which eventually leads to poor interparticle bonding.

The area of excellent consolidation may be identified by the production of components by LPBF that have a satisfactory degree of density and range from nearly 60% to 80%. Because of the energy density range that is used in this area, there is a rise in the temperature of the powder bed and a reduction in the viscosity of the melt pool. This results in an improvement in the densification of the fabricated item. This outcome was accomplished by using a greater energy density, which resulted in the production of molten powder [205]. The use of this energy density allowed for the achievement of the desired result.

The area of excessive balling may be identified by the presence of components that were manufactured using LPBF and have completely dense components but a highly rough surface. This outcome was accomplished by using a high laser energy density while maintaining a low scanning rate. These results were also validated by Zhang et al. [200]. They reported that the high percentage of liquid phase's formation may lead to balling.

7.5   LASER TECHNOLOGY FOR POST-TREATMENT OF COATINGS

Laser re-melting (LR) is a method that may be used to execute a thermal treatment of the coating surface in order to smooth out and close any open pores that may be present. The coating surface is melted during the LR process, which is then immediately followed by a period of fast cooling. This occurs because the thermal energy from the laser beam is sufficiently high. In most cases, LR is followed by the formation of a melting pool, which is controlled by the forces of buoyancy as a result of differences in density and convection as a result of gradients in surface tension [206–209]. Once the laser beam has passed through the area, the material underneath the molten layer serves as a heat sink, and solidification process begins from the solder–liquid cross-section toward the surface [206]. During the process of solidification, the linked pores in cold-sprayed coatings potentially decreased, and the splat structure has the potential to be eliminated. Therefore, LR is an auspicious approach for heat treatment of coating surfaces because it promotes a homogenous morphology, which has the potential to enhance better mechanical, tribological, and corrosive-resistant properties [210].

Cold-sprayed pure Ti coatings were modified by LR in previous studies [211,212]. Investigators found that the change eliminated inter-particulate porosity in the coating's outermost layer [211,212]. LR was applied (Figure 7.8a). The LRed area became denser as a consequence of its melting, which led to the subsequent solidification of the material and the release of micropores onto the free surface. In addition, the LRed Ti layer was accountable for the development of three characteristic metallurgical structural zones. These zones are denoted by the acronyms "re-melted zone," "heat-affected zone," and "base material," respectively (Figure 7.8). As a result of an impact at higher kinetic energy and significant deformation that occurred during the process of cold spray deposition, thin lamellae were detected in the BM zone. The HAZ experienced an increase in grain size as a direct consequence of particular heat input, which tends toward the formation of coarser lamellae. In between the BM and HAZ, a transition zone was developed (Figure 7.8d). As a result of the quick cooling that occurred during the LR process, the RZ developed equiaxial grains in addition to martensitic acicular grains. In a similar fashion, a transition zone was seen to exist between the RZ and the HAZ (Figure 7.8f), and inside this zone, some acicular martensitic grains were generated due to rapid rate of cooling.

The mechanical characteristics of the coatings would be affected, as a result, by the microstructural features that were shown in various zones [211]. Hardness in the HAZ was significantly lower than that measured in the BM because of the formation

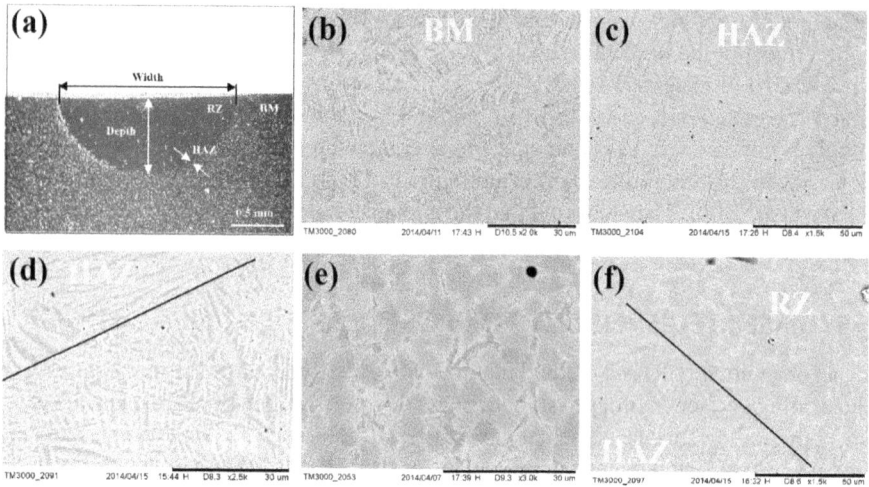

FIGURE 7.8 Optical microscopic image: (a) Cross-section of cold spray-deposited Ti coating after LR, and (b-f) SEM images at different locations: (b) BM, i.e., base material, (c) HAZ, (d) interface of BM and HAZ; (e) RZ, i.e., re-melted zone, and (f) interface of HAZ and RZ [211]. (Permissions under Creative common licence 4.0.)

of coarser lamellae there. It was discovered, however, that the RZ was harder than both the BM and the HAZ due to the presence of acicular martensitic grains within the RZ. In terms of wear and corrosion resistance, the upper surface oxide layer developed by LR was significantly more robust than that formed by BM. Oxides on the surface of the material might potentially affect its ductility as well as its capacity to resist fatigue and creep [213]. Researchers [211,212] conducted research on the influence that the speed of the laser scan had on cold-sprayed titanium coatings. While the laser scan speed was extremely low, the material evaporation took place, and surface fissures were generated. This was revealed to be the result of high heat input. In contrast, when the scan speed was excessively high owing to inadequate heat input, there was no significant alteration. In addition, Poza et al. [214] looked at how the mechanical characteristics of cold-sprayed Inconel 625 coatings were affected by the varied laser scan speeds. The findings demonstrated that increasing the scan speed resulted in a reduced heat input, which led to a drop in the LR layer depth. In addition, the coating porosity was dramatically decreased following the LR treatment, which led to an increase in the elastic modulus. Laser's scanning speed was increased, while the LRed coating's elastic elasticity remained unchanged (Figure 7.9). However, the LRed coating displayed softer than anticipated hardness because of the presence of columnar dendritic microstructure. This was because, following LR treatment, the highly deformed microstructure of the as-sprayed material relaxed. With a higher laser scan speed, the hardness drop was more noticeable because the Fe element diffused from the substrate into the coating at a slower scanning speed (Figure 7.9). This was attributed to the fact that when the laser's scanning speed was lower, the diffusion of Fe element occurred (more heat input). Therefore,

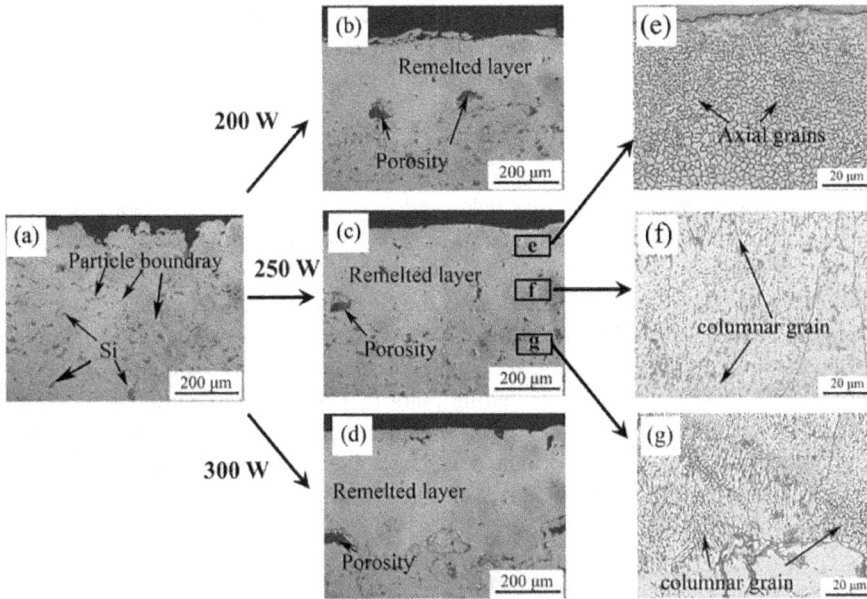

FIGURE 7.9 Cross-sectional SEM images: (a) as-coated Al-Si and (b) Al-Si coating after LR at 200W, (c) Al-Si coating after LR at 250W, (d) Al-Si coating after LR at 300W, and (e) top surface's morphology, (f) middle surface's morphology, and (g) bottom surface's morphology [215]. (Permission from ©Elsevier.)

LR has an effect not only on the coating layer itself but also on the interaction that occurs between the coating and the BM when there is adequate heat input.

The effects of laser power, in addition to laser scan speed, on the microstructure and mechanical properties of cold-sprayed Ti64 coatings were studied by Khun et al. [210]. It was found that the rapid cooling that resulted from using a stronger laser caused the surface to split and shrink along the lines that were irradiated. While the Ti64 coatings' increased surface hardness and wear resistance were a result of rapid cooling, the resulting surface fissures might have a negative impact on the properties of other materials. In order to create a microstructure and mechanical qualities that are acceptable, it is crucial to remember that the LR processing conditions—laser power, spot diameter, and traverse speed—should be adequately optimised to guarantee that the heat input and temperature distribution can be regulated.

In addition, the LRed coating demonstrated a shift in the noble direction in the values of the open circuit potential (OCP) and corrosion potential (E_{corr}), as well as a considerable drop in the potentiodynamic polarisation current (I_{pp}) [213]. These results suggest that the LRed coating is superior to the as-sprayed coating in terms of protecting against corrosion. It is possible that the LRed coating's high performance is the result of the combination of its thick LR layer and the surface oxide layer that protected it. Although the LRed Ti coating exhibits corrosion behaviour unique from that of the bulk Ti, the full extent of the LRed coating layer's corrosion process remains obscure. Thus, it is advised that additional research be conducted on the topic. More attention should be paid, in particular, to the one-of-a-kind microstructural characteristics as well as the function of the surface oxide coating on the LRed layer.

Kang et al. [215] investigated the impact of LR on cold-sprayed coatings composed of Al-Si composites, as opposed to studies that focused on coatings formed of pure metal or alloys. The Si particles and the Al matrix in the as-sprayed Al-Si coating interacted little, if at all, due to the mechanical bonding nature of the connection between these two phases [215]. Figure 7.9a depicts the as-sprayed deposit's microstructure was much improved, and the Si particles were completely invisible inside the LRed layer. The reason for this is because a Si-containing supersaturated Al phase was created by melting the Al-Si combination (Al has a melting temperature of 660°C, whereas Si has a melting point of 1414°C) and then quickly cooling it [215]. The creation of a Si-containing supersaturated Al phase has the potential to dramatically improve the performance of an Al-Si coating via lattice distortion and grain refinement. This phenomenon may give rise to a new research topic regarding the combination of LR and cold-spraying hybrid materials to produce new phases in the coating. The particles that make up hybrid materials vary. Additionally, equiaxed grains appeared in the upper area; columnar grains, which constituted the bulk of the grain structure, were observed in the lower region; and a combination of equiaxed and vertically columnar grains was observed in between. This was observed in the region that was in between the top and bottom regions (Figure 7.9e–g). Large holes have recently appeared between the LRed layer and the rest of the cold-sprayed deposit, and this may be due to the Kirkendall effect [215,216] and the agglomerates of the smaller pores (Figure 7.9b–d). This may have a negative impact not only on the integrity of the coating but also on its physicochemical properties. Therefore, additional work that is more specific can be done to figure out how to get rid of the large pores that were formed in the zone between the LRed layer and the layer that is still present.

7.6 CONCLUDING REMARKS AND FUTURE RESEARCH DIRECTIONS

The knowledge of the LPBF process, as well as any current upgrades to it, is the subject of this study. LPBF has evolved into a versatile technology that can be used in a wide variety of metals and the alloys that they form; as a result, it is garnering a large amount of attention. After conducting an exhaustive investigation of the LPBF procedure, a number of essential takeaways have surfaced, each of which is of the utmost significance. In order to reduce the number of flaws in the product as much as possible, consideration is given to the relevance of the many process factors. The LPBF process may be used to any material, and the most essential thing to do in order to obtain the maximum possible density and the requisite mechanical behaviour on the LPBFed component is to accurately monitor the processing parameters. This will allow you to accomplish both of those goals. The application of the LPBF technique to metals and alloys has resolved many of the issues that were previously associated with the conventional production procedures.

The attributes that were inherited by the sample that was LPBFed demonstrate that LPBF is capable of producing samples with qualities that are superior to those generated by traditional approaches. The primary factor that determines how metals behave during the densification process is the fluctuation in laser energy density, which is both regulated and affected by a number of other process factors. The density increase may be traced back to a shift in the laser energy density that was brought on by changes to the various process parameters. Additionally, the influence of powder particle size and distribution is less significant in LPBF due to the fact that all particles undergo full melting. In contrast to SLS, which involves the occurrence of partial melting, the powder parameters make a small attribution to the densification of the component. The top layer of cold-sprayed coatings may be densified by LR, which is great for wear and corrosion protection. More basic study is necessary; however, there are large holes between the re-melted layer and the underlying material. Excellent possibilities for cold-sprayed deposits to meet industrial standards may be found in the subsequent procedures.

REFERENCES

[1] Vasudev, H.; Singh, G.; Bansal, A.; Vardhan, S.; Thakur, L. Microwave Heating and Its Applications in Surface Engineering: A Review. *Mater. Res. Express*, **2019**, *6* (10), 102001. https://doi.org/10.1088/2053-1591/ab3674.

[2] Prashar, G.; Vasudev, H. A Comprehensive Review on Sustainable Cold Spray Additive Manufacturing: State of the Art, Challenges and Future Challenges. *J. Clean. Prod.*, **2021**, *310* (April), 127606. https://doi.org/10.1016/j.jclepro.2021.127606.

[3] Singh, J. A Review on Mechanisms and Testing of Wear in Slurry Pumps, Pipeline Circuits and Hydraulic Turbines. *J. Tribol.*, **2021**, *143* (September), 1–83. https://doi.org/10.1115/1.4050977.

[4] Singh, J.; Singh, S. A Review on Machine Learning Aspect in Physics and Mechanics of Glasses. *Mater. Sci. Eng. B*, **2022**, *284* (July), 115858. https://doi.org/10.1016/j.mseb.2022.115858.

[5] Prashar, G.; Vasudev, H.; Thakur, L. Performance of Different Coating Materials against Slurry Erosion Failure in Hydrodynamic Turbines: A Review. *Eng. Fail. Anal.*, **2020**, *115* (May), 104622. https://doi.org/10.1016/j.engfailanal.2020.104622.

[6] Singh, J.; Kumar, S.; Mohapatra, S. Study on Role of Particle Shape in Erosion Wear of Austenitic Steel Using Image Processing Analysis Technique. *Proc. Inst. Mech. Eng. Part J J. Eng. Tribol.*, **2019**, *233* (5), 712–725. https://doi.org/10.1177/1350650118794698.

[7] Kumar, P.; Kumar, S.; Singh, J. Rheological and Computational Analysis of Crude Oil Transportation. *Int. J. Mech. Aerospace, Ind. Mechatron. Manuf. Eng.*, **2017**, *11* (2), 429–432.

[8] Singh, J. *Investigation on Slurry Erosion of Different Pumping Materials and Coatings*; Thapar Institute of Engineering and Technology, Patiala, India, 2019.

[9] Singh, J.; Singh, R.; Singh, S.; Vasudev, H.; Kumar, S. Reducing Scrap Due to Missed Operations and Machining Defects in 90PS Pistons. *Int. J. Interact. Des. Manuf.*, **2022**. https://doi.org/10.1007/s12008-022-01071-0.

[10] Singh, J.; Kumar, M.; Kumar, S.; Mohapatra, S. K. Properties of Glass-Fiber Hybrid Composites: A Review. *Polym. Plast. Technol. Eng.*, **2017**, *56* (5), 455–469. https://doi.org/10.1080/03602559.2016.1233271.

[11] Singh, S.; Garg, J.; Singh, P.; Singh, G.; Kumar, K. Effect of Hard Faced Cr-Alloy on Abrasive Wear of Low Carbon Rotavator Blades Using Design of Experiments. *Mater. Today Proc.*, **2018**, *5* (2), 3390–3395. https://doi.org/10.1016/j.matpr.2017.11.583.

[12] Singh, J.; Singh, J. P.; Singh, M.; Szala, M. Computational Analysis of Solid Particle-Erosion Produced by Bottom Ash Slurry in 90° Elbow. In *MATEC Web of Conferences*, vol. 252, p. 04008. EDP Sciences, 2019. https://doi.org/0.1051/matecconf/201925204008.

[13] Kumar, S.; Singh, J.; Mohapatra, S. K. Role of Particle Size in Assessment of Physico-Chemical Properties and Trace Elements of Indian Fly Ash. *Waste Manag. Res. J. a Sustain. Circ. Econ.*, **2018**, *36* (11), 1016–1022. https://doi.org/10.1177/0734242X18804033.

[14] Kumar, P.; Singh, J.; Singh, S. Neural Network Supported Flow Characteristics Analysis of Heavy Sour Crude Oil Emulsified by Ecofriendly Bio-Surfactant Utilized as a Replacement of Sweet Crude Oil. *Chem. Eng. J. Adv.*, **2022**, *11* (June), 100342. https://doi.org/10.1016/j.ceja.2022.100342.

[15] Singh, J.; Singh, J. P. Performance Analysis of Erosion Resistant Mo2C Reinforced WC-CoCr Coating for Pump Impeller with Taguchi's Method. *Ind. Lubr. Tribol.*, **2022**, *74* (4), 431–441. https://doi.org/10.1108/ILT-05-2020-0155.

[16] Singh, J. Slurry Erosion Performance Analysis and Characterization of High-Velocity Oxy-Fuel Sprayed Ni and Co Hardsurfacing Alloy Coatings. *J. King Saud Univ. - Eng. Sci.*, **2021**. https://doi.org/10.1016/j.jksues.2021.06.009.

[17] Singh, J. Application of Thermal Spray Coatings for Protection against Erosion, Abrasion, and Corrosion in Hydropower Plants and Offshore Industry. In *Thermal Spray Coatings*; Thakur, L., Vasudev, H., Eds.; CRC Press: Boca Raton, 2021; pp. 243–283. https://doi.org/10.1201/9781003213185-10.

[18] Singh, J. Tribo-Performance Analysis of HVOF Sprayed 86WC-10Co4Cr & Ni-Cr$_2$O$_3$ on AISI 316L Steel Using DOE-ANN Methodology. *Ind. Lubr. Tribol.*, **2021**, *73* (5), 727–735. https://doi.org/10.1108/ILT-04-2020-0147.

[19] Singh, J. Analysis on Suitability of HVOF Sprayed Ni-20Al, Ni-20Cr and Al-20Ti Coatings in Coal-Ash Slurry Conditions Using Artificial Neural Network Model. *Ind. Lubr. Tribol.*, **2019**, *71* (7), 972–982. https://doi.org/10.1108/ILT-12-2018-0460.

[20] Singh, J.; Kumar, S.; Mohapatra, S. K. Erosion Wear Performance of Ni-Cr-O and NiCrBSiFe-WC(Co) Composite Coatings Deposited by HVOF Technique. *Ind. Lubr. Tribol.*, **2019**, *71* (4), 610–619. https://doi.org/10.1108/ILT-04-2018-0149.

[21] Singh, J.; Singh, S.; Gill, R. Applications of Biopolymer Coatings in Biomedical Engineering. *J. Electrochem. Sci. Eng.*, **2022**. https://doi.org/10.5599/jese.1460.

[22] Singh, J.; Singh, S. Neural Network Supported Study on Erosive Wear Performance Analysis of Y$_2$O$_3$/WC-10Co4Cr HVOF Coating. *J. King Saud Univ. - Eng. Sci.*, **2022**. https://doi.org/10.1016/j.jksues.2021.12.005.

[23] Singh, J.; Singh, J. P. Numerical Analysis on Solid Particle Erosion in Elbow of a Slurry Conveying Circuit. *J. Pipeline Syst. Eng. Pract.*, **2021**, *12* (1), 04020070. https://doi.org/10.1061/(asce)ps.1949-1204.0000518.

[24] Singh, J.; Mohapatra, S. K.; Kumar, S. Performance Analysis of Pump Materials Employed in Bottom Ash Slurry Erosion Conditions. *J. Tribol.*, **2021**, *30*, 73–89.

[25] Singh, J.; Kumar, S.; Mohapatra, S. K. Optimization of Erosion Wear Influencing Parameters of HVOF Sprayed Pumping Material for Coal-Water Slurry. *Mater. Today Proc.*, **2018**, *5* (11), 23789–23795. https://doi.org/10.1016/j.matpr.2018.10.170.

[26] Singh, J.; Gill, H. S.; Vasudev, H. Computational Fluid Dynamics Analysis on Effect of Particulate Properties on Erosive Degradation of Pipe Bends. *Int. J. Interact. Des. Manuf.*, **2022**. https://doi.org/10.1007/s12008-022-01094-7.

[27] Kumar, K.; Kumar, S.; Singh, G.; Singh, J.; Singh, J. Erosion Wear Investigation of HVOF Sprayed WC-10Co4Cr Coating on Slurry Pipeline Materials. *Coatings*, **2017**, *7* (4), 54. https://doi.org/10.3390/coatings7040054.

[28] Singh, J.; Kumar, S.; Singh, J. P.; Kumar, P.; Mohapatra, S. K. CFD Modeling of Erosion Wear in Pipe Bend for the Flow of Bottom Ash Suspension. *Part. Sci. Technol.*, **2019**, *37* (3), 275–285. https://doi.org/10.1080/02726351.2017.1364816.

[29] Singh, J.; Kumar, S.; Mohapatra, S. K. Erosion Tribo-Performance of HVOF Deposited Stellite-6 and Colmonoy-88 Micron Layers on SS-316L. *Tribol. Int.*, **2020**, *147* (June 2018), 105262. https://doi.org/10.1016/j.triboint.2018.06.004.

[30] Singh, J.; Singh, S. Neural Network Prediction of Slurry Erosion of Heavy-Duty Pump Impeller/Casing Materials 18Cr-8Ni, 16Cr-10Ni-2Mo, Super Duplex 24Cr-6Ni-3Mo-N, and Grey Cast Iron. *Wear*, **2021**, *476* (January), 203741. https://doi.org/10.1016/j.wear.2021.203741.

[31] Singh, J.; Kumar, S.; Mohapatra, S. K. Study on Solid Particle Erosion of Pump Materials by Fly Ash Slurry Using Taguchi's Orthogonal Array. *Tribol. - Finnish J. Tribol.*, **2021**, *38* (3–4), 31–38. https://doi.org/10.30678/fjt.97530.

[32] Kumar, P.; Kumar, S.; Singh, J. Effect of Natural Surfactant on the Rheological Characteristics of Heavy Crude Oil. *Mater. Today Proc.*, **2018**, *5* (11), 23881–23887. https://doi.org/10.1016/j.matpr.2018.10.180.

[33] Kumar, P.; Singh, J. Computational Study on Effect of Mahua Natural Surfactant on the Flow Properties of Heavy Crude Oil in a 90° Bend. *Mater. Today Proc.*, **2021**, *43*, 682–688. https://doi.org/10.1016/j.matpr.2020.12.612.

[34] Kumar, S.; Singh, M.; Singh, J.; Singh, J. P.; Kumar, S. Rheological Characteristics of Uni/Bi-Variant Particulate Iron Ore Slurry: Artificial Neural Network Approach. *J. Min. Sci.*, **2019**, *55* (2), 201–212. https://doi.org/10.1134/S1062739119025468.

[35] Szala, M.; Łatka, L.; Walczak, M.; Winnicki, M. Comparative Study on the Cavitation Erosion and Sliding Wear of Cold-Sprayed Al/Al$_2$O$_3$ and Cu/Al$_2$O$_3$ Coatings, and Stainless Steel, Aluminium Alloy, Copper and Brass. *Metals*, **2020**, *10* (856), 1–24. https://doi.org/10.3390/met10070856.

[36] Szala, M.; Łukasik, D. Cavitation Wear of Pump Impellers. *J. Technol. Exploit. Mech. Eng.*, **2016**, *2* (1), 40–44. https://doi.org/10.35784/jteme.337.

[37] Szala, M.; Hejwowski, T. J.; Lenart, I. Cavitation Erosion Resistance of Ni-Co Based Coatings. *Adv. Sci. Technol. Res. J.*, **2014**, *8* (21), 36–42. https://doi.org/10.12913/22998624.1091876.

[38] Singh, J. Wear Performance Analysis and Characterization of HVOF Deposited Ni–20Cr$_2$O$_3$, Ni–30Al$_2$O$_3$, and Al$_2$O$_3$–13TiO$_2$ Coatings. *Appl. Surf. Sci. Adv.*, **2021**, *6*, 100161. https://doi.org/10.1016/j.apsadv.2021.100161.

[39] Singh, J.; Kumar, S.; Mohapatra, S. K. Tribological Performance of Yttrium (III) and Zirconium (IV) Ceramics Reinforced WC–10Co4Cr Cermet Powder HVOF Thermally Sprayed on X2CrNiMo-17-12-2 Steel. *Ceram. Int.*, **2019**, *45* (17), 23126–23142. https://doi.org/10.1016/j.ceramint.2019.08.007.

[40] Singh, J.; Singh, S.; Pal Singh, J. Investigation on Wall Thickness Reduction of Hydropower Pipeline Underwent to Erosion-Corrosion Process. *Eng. Fail. Anal.*, **2021**, *127* (May), 105504. https://doi.org/10.1016/j.engfailanal.2021.105504.

[41] Singh, J.; Kumar, S.; Mohapatra, S. K. Tribological Analysis of WC–10Co–4Cr and Ni–20Cr$_2$O$_3$ Coating on Stainless Steel 304. *Wear*, **2017**, *376–377*, 1105–1111. https://doi.org/10.1016/j.wear.2017.01.032.

[42] Singh, J.; Kumar, S.; Mohapatra, S. K.; Kumar, S. Shape Simulation of Solid Particles by Digital Interpretations of Scanning Electron Micrographs Using IPA Technique. *Mater. Today Proc.*, **2018**, *5* (9), 17786–17791. https://doi.org/10.1016/j.matpr.2018.06.103.

[43] Singh, J.; Kumar, S.; Singh, G. Taguchi's Approach for Optimization of Tribo-Resistance Parameters Forss304. *Mater. Today Proc.*, **2018**, *5* (2), 5031–5038. https://doi.org/10.1016/j.matpr.2017.12.081.

[44] Sunitha, K.; Vasudev, H. Microsrtructural and Mechanical Characterization of HVOF-Sprayed Ni-Based Alloy Coating. *Int. J. Surf. Eng. Interdiscip. Mater. Sci.*, **2022**, *10* (1), 1–9. https://doi.org/10.4018/IJSEIMS.298705.

[45] Sharma, Y.; Singh, K. J.; Vasudev, H. Experimental Studies on Friction Stir Welding of Aluminium Alloys. *Mater. Today Proc.*, **2022**, *50*, 2387–2391. https://doi.org/10.1016/j.matpr.2021.10.254.

[46] Sunitha, K.; Vasudev, H. A Short Note on the Various Thermal Spray Coating Processes and Effect of Post-Treatment on Ni-Based Coatings. *Mater. Today Proc.*, **2022**, *50*, 1452–1457. https://doi.org/10.1016/j.matpr.2021.09.017.

[47] Parkash, J.; Saggu, H. S.; Vasudev, H. A Short Review on the Performance of High Velocity Oxy-Fuel Coatings in Boiler Steel Applications. *Mater. Today Proc.*, **2022**, *50*, 1442–1446. https://doi.org/10.1016/j.matpr.2021.09.014.

[48] Satyavathi Yedida, V. V.; Vasudev, H. A Review on the Development of Thermal Barrier Coatings by Using Thermal Spray Techniques. *Mater. Today Proc.*, **2022**, *50*, 1458–1464. https://doi.org/10.1016/j.matpr.2021.09.018.

[49] Prashar, G.; Vasudev, H. Surface Topology Analysis of Plasma Sprayed Inconel625-Al_2O_3 Composite Coating. *Mater. Today Proc.*, **2022**, *50*, 607–611. https://doi.org/10.1016/j.matpr.2021.03.090.

[50] Prashar, G.; Vasudev, H. Hot Corrosion Behavior of Super Alloys. *Mater. Today Proc.*, **2020**, *26*, 1131–1135. https://doi.org/10.1016/j.matpr.2020.02.226.

[51] Ganesh Reddy Majji, B.; Vasudev, H.; Bansal, A. A Review on the Oxidation and Wear Behavior of the Thermally Sprayed High-Entropy Alloys. *Mater. Today Proc.*, **2022**, *50*, 1447–1451. https://doi.org/10.1016/j.matpr.2021.09.016.

[52] Singh, J.; Vasudev, H.; Singh, S. Performance of Different Coating Materials against High Temperature Oxidation in Boiler Tubes – A Review. *Mater. Today Proc.*, **2020**, *26*, 972–978. https://doi.org/10.1016/j.matpr.2020.01.156.

[53] Mehta, A.; Vasudev, H.; Singh, S. Recent Developments in the Designing of Deposition of Thermal Barrier Coatings – A Review. *Mater. Today Proc.*, **2020**, *26*, 1336–1342. https://doi.org/10.1016/j.matpr.2020.02.271.

[54] Singh, M.; Vasudev, H.; Kumar, R. Microstructural Characterization of BN Thin Films Using RF Magnetron Sputtering Method. *Mater. Today Proc.*, **2020**, *26*, 2277–2282. https://doi.org/10.1016/j.matpr.2020.02.493.

[55] Vasudev, H.; Thakur, L.; Singh, H.; Bansal, A. Erosion Behaviour of HVOF Sprayed Alloy718-Nano Al_2O_3 Composite Coatings on Grey Cast Iron at Elevated Temperature Conditions. *Surf. Topogr. Metrol. Prop.*, **2021**, *9* (3), 035022. https://doi.org/10.1088/2051-672X/ac1c80.

[56] Prashar, G.; Vasudev, H.; Thakur, L. Influence of Heat Treatment on Surface Properties of HVOF Deposited WC and Ni-Based Powder Coatings: A Review. *Surf. Topogr. Metrol. Prop.*, **2021**, *9* (4), 043002. https://doi.org/10.1088/2051-672X/ac3a52.

[57] Vasudev, H.; Prashar, G.; Thakur, L.; Bansal, A. Electrochemical Corrosion Behavior and Microstructural Characterization of HVOF Sprayed Inconel-718 Coating on Gray Cast Iron. *J. Fail. Anal. Prev.*, **2021**, *21* (1), 250–260. https://doi.org/10.1007/s11668-020-01057-8.

[58] Vasudev, H.; Prashar, G.; Thakur, L.; Bansal, A. Electrochemical Corrosion Behavior and Microstructural Characterization of Hvof Sprayed Inconel718-Al_2O_3 Composite Coatings. *Surf. Rev. Lett.*, **2022**, *29* (02), 2250017. https://doi.org/10.1142/S0218625X22500172.

[59] Vasudev, H.; Thakur, L.; Singh, H.; Bansal, A. A Study on Processing and Hot Corrosion Behaviour of HVOF Sprayed Inconel718-Nano Al_2O_3 Coatings. *Mater. Today Commun.*, **2020**, *25* (May), 101626. https://doi.org/10.1016/j.mtcomm.2020.101626.

[60] Vasudev, H.; Prashar, G.; Thakur, L.; Bansal, A. Microstructural Characterization and Electrochemical Corrosion Behaviour of HVOF Sprayed Alloy718-NanoAl_2O_3 Composite Coatings. *Surf. Topogr. Metrol. Prop.*, **2021**, *9* (3), 035003. https://doi.org/10.1088/2051-672X/ac1044.

[61] Vasudev, H.; Thakur, L.; Bansal, A.; Singh, H.; Zafar, S. High Temperature Oxidation and Erosion Behaviour of HVOF Sprayed Bi-Layer Alloy-718/NiCrAlY Coating. *Surf. Coatings Technol.*, **2019**, *362* (May 2018), 366–380. https://doi.org/10.1016/j.surfcoat.2019.02.012.

[62] Prashar, G.; Vasudev, H.; Thakur, L. High-Temperature Oxidation and Erosion Resistance of Ni-Based Thermally-Sprayed Coatings Used In Power Generation Machinery: A Review. *Surf. Rev. Lett.*, **2022**, *29* (03), 2230003. https://doi.org/10.1142/S0218625X22300039.

[63] Prashar, G.; Vasudev, H. Structure–Property Correlation of Plasma-Sprayed Inconel625-Al$_2$O$_3$ Bimodal Composite Coatings for High-Temperature Oxidation Protection. *J. Therm. Spray Technol.*, **2022**, *31* (8), 2385–2408. https://doi.org/10.1007/s11666-022-01466-1.

[64] Prashar, G.; Vasudev, H. High Temperature Erosion Behavior of Plasma Sprayed Al$_2$O$_3$ Coating on AISI-304 Stainless Steel. *World J. Eng.*, **2021**, *18* (5), 760–766. https://doi.org/10.1108/WJE-10-2020-0476.

[65] Prashar, G.; Vasudev, H. Structure-Property Correlation and High-Temperature Erosion Performance of Inconel625-Al$_2$O$_3$ Plasma-Sprayed Bimodal Composite Coatings. *Surf. Coatings Technol.*, **2022**, *439* (April), 128450. https://doi.org/10.1016/j.surfcoat.2022.128450.

[66] Mudgal, D.; Ahuja, L.; Bhatia, D.; Singh, S.; Prakash, S. High Temperature Corrosion Behaviour of Superalloys under Actual Waste Incinerator Environment. *Eng. Fail. Anal.*, **2016**, *63*, 160–171. https://doi.org/10.1016/j.engfailanal.2016.02.016.

[67] Singh, P.; Vasudev, H.; Bansal, A. Effect of Post-Heat Treatment on the Microstructural, Mechanical, and Bioactivity Behavior of the Microwave-Assisted Alumina-Reinforced Hydroxyapatite Cladding. *Proc. Inst. Mech. Eng. Part E J. Process Mech. Eng.*, **2022**, 095440892211161. https://doi.org/10.1177/09544089221116168.

[68] Singh, M.; Vasudev, H.; Kumar, R. Corrosion and Tribological Behaviour of BN Thin Films Deposited Using Magnetron Sputtering. *Int. J. Surf. Eng. Interdiscip. Mater. Sci.*, **2021**, *9* (2), 24–39. https://doi.org/10.4018/IJSEIMS.2021070102.

[69] Singh, M.; Vasudev, H.; Singh, M. Surface Protection of SS-316L with Boron Nitride Based Thin Films Using Radio Frequency Magnetron Sputtering Technique. *J. Electrochem. Sci. Eng.*, **2022**, *12* (5), 851–863. https://doi.org/10.5599/jese.1247.

[70] Singh, J.; Singh, S.; Verma, A. Artificial Intelligence in Use of ZrO2 Material in Biomedical Science. *J. Electrochem. Sci. Eng.*, **2022**. https://doi.org/10.5599/jese.1498.

[71] Kumar, D.; Yadav, R.; Singh, J. Evolution and Adoption of Microwave Claddings in Modern Engineering Applications. In *Advances in Microwave Processing for Engineering Materials*; CRC Press: Boca Raton, 2022; pp. 134–153. https://doi.org/10.1201/9781003248743-8.

[72] Singh, R.; Toseef, M.; Kumar, J.; Singh, J. Benefits and Challenges in Additive Manufacturing and Its Applications. In *Sustainable Advanced Manufacturing and Materials Processing*; Kaushal, S., Singh, I., Singh, S., Gupta, A., Eds.; CRC Press: Boca Raton, 2022; pp. 137–157. https://doi.org/10.1201/9781003269298-8.

[73] Dykhuizen, R. C.; Smith, M. F. Gas Dynamic Principles of Cold Spray. *J. Therm. Spray Technol.*, **1998**, *7* (2), 205–212. https://doi.org/10.1361/105996398770350945.

[74] Moridi, A.; Hassani-Gangaraj, S. M.; Guagliano, M.; Dao, M. Cold Spray Coating: Review of Material Systems and Future Perspectives. *Surf. Eng.*, **2014**, *30* (6), 369–395. https://doi.org/10.1179/1743294414Y.0000000270.

[75] Yin, S.; Meyer, M.; Li, W.; Liao, H.; Lupoi, R. Gas Flow, Particle Acceleration, and Heat Transfer in Cold Spray: A Review. *J. Therm. Spray Technol.*, **2016**, *25* (5), 874–896. https://doi.org/10.1007/s11666-016-0406-8.

[76] Yang, K.; Li, W.; Yang, X.; Xu, Y. Anisotropic Response of Cold Sprayed Copper Deposits. *Surf. Coatings Technol.*, **2018**, *335*, 219–227. https://doi.org/10.1016/j.surfcoat.2017.12.043.

[77] Chen, C.; Gojon, S.; Xie, Y.; Yin, S.; Verdy, C.; Ren, Z.; Liao, H.; Deng, S. A Novel Spiral Trajectory for Damage Component Recovery with Cold Spray. *Surf. Coatings Technol.*, **2017**, *309*, 719–728. https://doi.org/10.1016/j.surfcoat.2016.10.096.

[78] Cai, Z.; Chen, T.; Zeng, C.; Guo, X.; Lian, H.; Zheng, Y.; Wei, X. A Global Approach to the Optimal Trajectory Based on an Improved Ant Colony Algorithm for Cold Spray. *J. Therm. Spray Technol.*, **2016**, *25* (8), 1631–1637. https://doi.org/10.1007/s11666-016-0468-7.

[79] Herzog, D.; Seyda, V.; Wycisk, E.; Emmelmann, C. Additive Manufacturing of Metals. *Acta Mater.*, **2016**, *117*, 371–392. https://doi.org/10.1016/j.actamat.2016.07.019.

[80] Frazier, W. E. Metal Additive Manufacturing: A Review. *J. Mater. Eng. Perform.*, **2014**, *23* (6), 1917–1928. https://doi.org/10.1007/s11665-014-0958-z.

[81] Shin, Y. C.; Bailey, N.; Katinas, C.; Tan, W. Predictive Modeling Capabilities from Incident Powder and Laser to Mechanical Properties for Laser Directed Energy Deposition. *Comput. Mech.*, **2018**, *61* (5), 617–636. https://doi.org/10.1007/s00466-018-1545-1.

[82] Ahsan, F.; Ladani, L. Temperature Profile, Bead Geometry, and Elemental Evaporation in Laser Powder Bed Fusion Additive Manufacturing Process. *JOM*, **2020**, *72* (1), 429–439. https://doi.org/10.1007/s11837-019-03872-3.

[83] Liu, S.; Shin, Y. C. Simulation and Experimental Studies on Microstructure Evolution of Resolidified Dendritic TiC in Laser Direct Deposited Ti-TiC Composite. *Mater. Des.*, **2018**, *159*, 212–223. https://doi.org/10.1016/j.matdes.2018.08.053.

[84] Parry, L.; Ashcroft, I. A.; Wildman, R. D. Understanding the Effect of Laser Scan Strategy on Residual Stress in Selective Laser Melting through Thermo-Mechanical Simulation. *Addit. Manuf.*, **2016**, *12*, 1–15. https://doi.org/10.1016/j.addma.2016.05.014.

[85] Tammas-Williams, S.; Zhao, H.; Léonard, F.; Derguti, F.; Todd, I.; Prangnell, P. B. XCT Analysis of the Influence of Melt Strategies on Defect Population in Ti–6Al–4V Components Manufactured by Selective Electron Beam Melting. *Mater. Charact.*, **2015**, *102*, 47–61. https://doi.org/10.1016/j.matchar.2015.02.008.

[86] Strantza, M.; Vafadari, R.; de Baere, D.; Vrancken, B.; van Paepegem, W.; Vandendael, I.; Terryn, H.; Guillaume, P.; van Hemelrijck, D. Fatigue of Ti6Al4V Structural Health Monitoring Systems Produced by Selective Laser Melting. *Materials*, **2016**, *9* (2), 106. https://doi.org/10.3390/ma9020106.

[87] Biswas, N.; Ding, J. L.; Balla, V. K.; Field, D. P.; Bandyopadhyay, A. Deformation and Fracture Behavior of Laser Processed Dense and Porous Ti6Al4V Alloy under Static and Dynamic Loading. *Mater. Sci. Eng. A*, **2012**, *549*, 213–221. https://doi.org/10.1016/j.msea.2012.04.036.

[88] Beretta, S.; Romano, S. A Comparison of Fatigue Strength Sensitivity to Defects for Materials Manufactured by AM or Traditional Processes. *Int. J. Fatigue*, **2017**, *94*, 178–191. https://doi.org/10.1016/j.ijfatigue.2016.06.020.

[89] Tofail, S. A. M.; Koumoulos, E. P.; Bandyopadhyay, A.; Bose, S.; O'Donoghue, L.; Charitidis, C. Additive Manufacturing: Scientific and Technological Challenges, Market Uptake and Opportunities. *Mater. Today*, **2018**, *21* (1), 22–37. https://doi.org/10.1016/j.mattod.2017.07.001.

[90] Molotnikov, A.; Kingsbury, A.; Brandt, M. Current State and Future Trends in Laser Powder Bed Fusion Technology. In *Fundamentals of Laser Powder Bed Fusion of Metals*; Elsevier, 2021; pp. 621–634. https://doi.org/10.1016/B978-0-12-824090-8.00011-1.

[91] Wischeropp, T. M.; Tarhini, H.; Emmelmann, C. Influence of Laser Beam Profile on the Selective Laser Melting Process of AlSi10Mg. *J. Laser Appl.*, **2020**, *32* (2), 022059. https://doi.org/10.2351/7.0000100.

[92] Zhang, L.-C.; Attar, H. Selective Laser Melting of Titanium Alloys and Titanium Matrix Composites for Biomedical Applications: A Review. *Adv. Eng. Mater.*, **2016**, *18* (4), 463–475. https://doi.org/10.1002/adem.201500419.

[93] Kruth, J.; Mercelis, P.; Van Vaerenbergh, J.; Froyen, L.; Rombouts, M. Binding Mechanisms in Selective Laser Sintering and Selective Laser Melting. *Rapid Prototyp. J.*, **2005**, *11* (1), 26–36. https://doi.org/10.1108/13552540510573365.

[94] Riza, S. H.; Masood, S. H.; Wen, C. Laser-Assisted Additive Manufacturing for Metallic Biomedical Scaffolds. In *Comprehensive Materials Processing*; Elsevier, 2014; pp. 285–301. https://doi.org/10.1016/B978-0-08-096532-1.01017-7.

[95] Yadroitsev, I.; Smurov, I. Surface Morphology in Selective Laser Melting of Metal Powders. *Phys. Procedia*, **2011**, *12*, 264–270. https://doi.org/10.1016/j.phpro.2011.03.034.

[96] Yadroitsev, I.; Yadroitsava, I.; Smurov, I. Strategy of Fabrication of Complex Shape Parts Based on the Stability of Single Laser Melted Track. *Proc. SPIE*, **2011**, *7921*, 79210C. https://doi.org/10.1117/12.875402.

[97] Li, Y.; Gu, D. Parametric Analysis of Thermal Behavior during Selective Laser Melting Additive Manufacturing of Aluminum Alloy Powder. *Mater. Des.*, **2014**, *63*, 856–867. https://doi.org/10.1016/j.matdes.2014.07.006.

[98] Zhang, H. J.; Zhang, D. F.; Ma, C. H.; Guo, S. F. Improving Mechanical Properties and Corrosion Resistance of Mg-6Zn-Mn Magnesium Alloy by Rapid Solidification. *Mater. Lett.*, **2013**, *92*, 45–48. https://doi.org/10.1016/j.matlet.2012.10.051.

[99] Kruth, J.-P.; Levy, G.; Klocke, F.; Childs, T. H. C. Consolidation Phenomena in Laser and Powder-Bed Based Layered Manufacturing. *CIRP Ann.*, **2007**, *56* (2), 730–759. https://doi.org/10.1016/j.cirp.2007.10.004.

[100] Yap, C. Y.; Chua, C. K.; Dong, Z. L.; Liu, Z. H.; Zhang, D. Q.; Loh, L. E.; Sing, S. L. Review of Selective Laser Melting: Materials and Applications. *Appl. Phys. Rev.*, **2015**, *2* (4), 041101. https://doi.org/10.1063/1.4935926.

[101] Gurrappa, I. Characterization of Titanium Alloy Ti-6Al-4V for Chemical, Marine and Industrial Applications. *Mater. Charact.*, **2003**, *51* (2–3), 131–139. https://doi.org/10.1016/j.matchar.2003.10.006.

[102] Vasudev, H.; Singh, P.; Thakur, L.; Bansal, A. Mechanical and Microstructural Characterization of Microwave Post Processed Alloy-718 Coating. *Mater. Res. Express*, **2020**, *6* (12), 1265f5. https://doi.org/10.1088/2053-1591/ab66fb.

[103] Geetha, M.; Singh, A. K.; Asokamani, R.; Gogia, A. K. Ti Based Biomaterials, the Ultimate Choice for Orthopaedic Implants – A Review. *Prog. Mater. Sci.*, **2009**, *54* (3), 397–425. https://doi.org/10.1016/j.pmatsci.2008.06.004.

[104] Gu, D.; Hagedorn, Y.-C.; Meiners, W.; Meng, G.; Batista, R. J. S.; Wissenbach, K.; Poprawe, R. Densification Behavior, Microstructure Evolution, and Wear Performance of Selective Laser Melting Processed Commercially Pure Titanium. *Acta Mater.*, **2012**, *60* (9), 3849–3860. https://doi.org/10.1016/j.actamat.2012.04.006.

[105] Wang, S. Q.; Liu, J. H.; Chen, D. L. Effect of Strain Rate and Temperature on Strain Hardening Behavior of a Dissimilar Joint between Ti–6Al–4V and Ti17 Alloys. *Mater. Des.*, **2014**, *56*, 174–184. https://doi.org/10.1016/j.matdes.2013.11.003.

[106] Banerjee, D.; Williams, J. C. Perspectives on Titanium Science and Technology. *Acta Mater.*, **2013**, *61* (3), 844–879. https://doi.org/10.1016/j.actamat.2012.10.043.

[107] Abdel-Hady Gepreel, M.; Niinomi, M. Biocompatibility of Ti-Alloys for Long-Term Implantation. *J. Mech. Behav. Biomed. Mater.*, **2013**, *20*, 407–415. https://doi.org/10.1016/j.jmbbm.2012.11.014.

[108] Attar, H.; Calin, M.; Zhang, L. C.; Scudino, S.; Eckert, J. Manufacture by Selective Laser Melting and Mechanical Behavior of Commercially Pure Titanium. *Mater. Sci. Eng. A*, **2014**, *593*, 170–177. https://doi.org/10.1016/j.msea.2013.11.038.

[109] Vrancken, B.; Thijs, L.; Kruth, J.-P.; Van Humbeeck, J. Microstructure and Mechanical Properties of a Novel β Titanium Metallic Composite by Selective Laser Melting. *Acta Mater.*, **2014**, *68*, 150–158. https://doi.org/10.1016/j.actamat.2014.01.018.

[110] Cain, V.; Thijs, L.; Van Humbeeck, J.; Van Hooreweder, B.; Knutsen, R. Crack Propagation and Fracture Toughness of Ti6Al4V Alloy Produced by Selective Laser Melting. *Addit. Manuf.*, **2015**, *5*, 68–76. https://doi.org/10.1016/j.addma.2014.12.006.

[111] Yang, C.; Zhao, Y. J.; Kang, L. M.; Li, D. D.; Zhang, W. W.; Zhang, L. C. High-Strength Silicon Brass Manufactured by Selective Laser Melting. *Mater. Lett.*, **2018**, *210*, 169–172. https://doi.org/10.1016/j.matlet.2017.09.011.

[112] Ren, Y. M.; Lin, X.; Fu, X.; Tan, H.; Chen, J.; Huang, W. D. Microstructure and Deformation Behavior of Ti-6Al-4V Alloy by High-Power Laser Solid Forming. *Acta Mater.*, **2017**, *132*, 82–95. https://doi.org/10.1016/j.actamat.2017.04.026.

[113] Yao, J.; Suo, T.; Zhang, S.; Zhao, F.; Wang, H.; Liu, J.; Chen, Y.; Li, Y. Influence of Heat-Treatment on the Dynamic Behavior of 3D Laser-Deposited Ti–6Al–4V Alloy. *Mater. Sci. Eng. A*, **2016**, *677*, 153–162. https://doi.org/10.1016/j.msea.2016.09.036.

[114] Davari, N.; Rostami, A.; Abbasi, S. M. Effects of Annealing Temperature and Quenching Medium on Microstructure, Mechanical Properties as Well as Fatigue Behavior of Ti-6Al-4V Alloy. *Mater. Sci. Eng. A*, **2017**, *683*, 1–8. https://doi.org/10.1016/j.msea.2016.11.095.

[115] Brailovski, V.; Prokoshkin, S.; Gauthier, M.; Inaekyan, K.; Dubinskiy, S.; Petrzhik, M.; Filonov, M. Bulk and Porous Metastable Beta Ti–Nb–Zr(Ta) Alloys for Biomedical Applications. *Mater. Sci. Eng. C*, **2011**, *31* (3), 643–657. https://doi.org/10.1016/j.msec.2010.12.008.

[116] Fukuda, A.; Takemoto, M.; Saito, T.; Fujibayashi, S.; Neo, M.; Yamaguchi, S.; Kizuki, T.; Matsushita, T.; Niinomi, M.; Kokubo, T.; et al. Bone Bonding Bioactivity of Ti Metal and Ti–Zr–Nb–Ta Alloys with Ca Ions Incorporated on Their Surfaces by Simple Chemical and Heat Treatments. *Acta Biomater.*, **2011**, *7* (3), 1379–1386. https://doi.org/10.1016/j.actbio.2010.09.026.

[117] Shukla, A. K.; Balasubramaniam, R. Effect of Surface Treatment on Electrochemical Behavior of CP Ti, Ti–6Al–4V and Ti–13Nb–13Zr Alloys in Simulated Human Body Fluid. *Corros. Sci.*, **2006**, *48* (7), 1696–1720. https://doi.org/10.1016/j.corsci.2005.06.003.

[118] Lin, C.-W.; Ju, C.-P.; Chern Lin, J.-H. A Comparison of the Fatigue Behavior of Cast Ti–7.5Mo with c.p. Titanium, Ti–6Al–4V and Ti–13Nb–13Zr Alloys. *Biomaterials*, **2005**, *26* (16), 2899–2907. https://doi.org/10.1016/j.biomaterials.2004.09.007.

[119] Cvijović-Alagić, I.; Cvijović, Z.; Mitrović, S.; Panić, V.; Rakin, M. Wear and Corrosion Behaviour of Ti–13Nb–13Zr and Ti–6Al–4V Alloys in Simulated Physiological Solution. *Corros. Sci.*, **2011**, *53* (2), 796–808. https://doi.org/10.1016/j.corsci.2010.11.014.

[120] Fischer, M.; Joguet, D.; Robin, G.; Peltier, L.; Laheurte, P. In Situ Elaboration of a Binary Ti–26Nb Alloy by Selective Laser Melting of Elemental Titanium and Niobium Mixed Powders. *Mater. Sci. Eng. C*, **2016**, *62*, 852–859. https://doi.org/10.1016/j.msec.2016.02.033.

[121] Zhou, L.; Yuan, T.; Li, R.; Tang, J.; Wang, M.; Mei, F. Microstructure and Mechanical Properties of Selective Laser Melted Biomaterial Ti-13Nb-13Zr Compared to Hot-Forging. *Mater. Sci. Eng. A*, **2018**, *725*, 329–340. https://doi.org/10.1016/j.msea.2018.04.001.

[122] Zhou, L.; Yuan, T.; Li, R.; Tang, J.; Wang, M.; Mei, F. Anisotropic Mechanical Behavior of Biomedical Ti-13Nb-13Zr Alloy Manufactured by Selective Laser Melting. *J. Alloys Compd.*, **2018**, *762*, 289–300. https://doi.org/10.1016/j.jallcom.2018.05.179.

[123] Dimić, I.; Cvijović-Alagić, I.; Völker, B.; Hohenwarter, A.; Pippan, R.; Veljović, Đ.; Rakin, M.; Bugarski, B. Microstructure and Metallic Ion Release of Pure Titanium and Ti–13Nb–13Zr Alloy Processed by High Pressure Torsion. *Mater. Des.*, **2016**, *91*, 340–347. https://doi.org/10.1016/j.matdes.2015.11.088.

[124] Mohan, L.; Anandan, C. Wear and Corrosion Behavior of Oxygen Implanted Biomedical Titanium Alloy Ti–13Nb–13Zr. *Appl. Surf. Sci.*, **2013**, *282*, 281–290. https://doi.org/10.1016/j.apsusc.2013.05.120.

[125] Khan, M. A.; Williams, R. L.; Williams, D. F. The Corrosion Behaviour of Ti–6Al–4V, Ti–6Al–7Nb and Ti–13Nb–13Zr in Protein Solutions. *Biomaterials*, **1999**, *20* (7), 631–637. https://doi.org/10.1016/S0142-9612(98)00217-8.

[126] Geetha, M.; Kamachi Mudali, U.; Gogia, A.; Asokamani, R.; Raj, B. Influence of Microstructure and Alloying Elements on Corrosion Behavior of Ti–13Nb–13Zr Alloy. *Corros. Sci.*, **2004**, *46* (4), 877–892. https://doi.org/10.1016/S0010-938X(03)00186-0.

[127] Urbańczyk, E.; Krząkała, A.; Kazek-Kęsik, A.; Michalska, J.; Stolarczyk, A.; Dercz, G.; Simka, W. Electrochemical Modification of Ti–13Nb–13Zr Alloy Surface in Phosphate Based Solutions. *Surf. Coatings Technol.*, **2016**, *291*, 79–88. https://doi.org/10.1016/j.surfcoat.2016.02.025.

[128] Bobbili, R.; Madhu, V. Dynamic Recrystallization Behavior of a Biomedical Ti–13Nb–13Zr Alloy. *J. Mech. Behav. Biomed. Mater.*, **2016**, *59*, 146–155. https://doi.org/10.1016/j.jmbbm.2015.12.025.

[129] Quan, G.; Zhang, L.; Wang, X.; Li, Y. Correspondence between Microstructural Evolution Mechanisms and Hot Processing Parameters for Ti-13Nb-13Zr Biomedical Alloy in Comprehensive Processing Maps. *J. Alloys Compd.*, **2017**, *698*, 178–193. https://doi.org/10.1016/j.jallcom.2016.12.140.

[130] Zhang, L.; Song, B.; Fu, J. J.; Wei, S. S.; Yang, L.; Yan, C. Z.; Li, H.; Gao, L.; Shi, Y. S. Topology-Optimized Lattice Structures with Simultaneously High Stiffness and Light Weight Fabricated by Selective Laser Melting: Design, Manufacturing and Characterization. *J. Manuf. Process.*, **2020**, *56*, 1166–1177. https://doi.org/10.1016/j.jmapro.2020.06.005.

[131] Wally, Z. J.; Haque, A. M.; Feteira, A.; Claeyssens, F.; Goodall, R.; Reilly, G. C. Selective Laser Melting Processed Ti6Al4V Lattices with Graded Porosities for Dental Applications. *J. Mech. Behav. Biomed. Mater.*, **2019**, *90*, 20–29. https://doi.org/10.1016/j.jmbbm.2018.08.047.

[132] Maskery, I.; Aremu, A. O.; Parry, L.; Wildman, R. D.; Tuck, C. J.; Ashcroft, I. A. Effective Design and Simulation of Surface-Based Lattice Structures Featuring Volume Fraction and Cell Type Grading. *Mater. Des.*, **2018**, *155*, 220–232. https://doi.org/10.1016/j.matdes.2018.05.058.

[133] Vrancken, B.; Thijs, L.; Kruth, J.-P.; Van Humbeeck, J. Heat Treatment of Ti6Al4V Produced by Selective Laser Melting: Microstructure and Mechanical Properties. *J. Alloys Compd.*, **2012**, *541*, 177–185. https://doi.org/10.1016/j.jallcom.2012.07.022.

[134] Xu, Y.; Zhang, D.; Hu, S.; Chen, R.; Gu, Y.; Kong, X.; Tao, J.; Jiang, Y. Mechanical Properties Tailoring of Topology Optimized and Selective Laser Melting Fabricated Ti6Al4V Lattice Structure. *J. Mech. Behav. Biomed. Mater.*, **2019**, *99*, 225–239. https://doi.org/10.1016/j.jmbbm.2019.06.021.

[135] Zhang, L.; Song, B.; Yang, L.; Shi, Y. Tailored Mechanical Response and Mass Transport Characteristic of Selective Laser Melted Porous Metallic Biomaterials for Bone Scaffolds. *Acta Biomater.*, **2020**, *112*, 298–315. https://doi.org/10.1016/j.actbio.2020.05.038.

[136] Chen, L. Y.; Huang, J. C.; Lin, C. H.; Pan, C. T.; Chen, S. Y.; Yang, T. L.; Lin, D. Y.; Lin, H. K.; Jang, J. S. C. Anisotropic Response of Ti-6Al-4V Alloy Fabricated by 3D Printing Selective Laser Melting. *Mater. Sci. Eng. A*, **2017**, *682*, 389–395. https://doi.org/10.1016/j.msea.2016.11.061.

[137] Qiu, C.; Adkins, N. J. E.; Attallah, M. M. Microstructure and Tensile Properties of Selectively Laser-Melted and of HIPed Laser-Melted Ti–6Al–4V. *Mater. Sci. Eng. A*, **2013**, *578*, 230–239. https://doi.org/10.1016/j.msea.2013.04.099.

[138] Shi, X.; Ma, S.; Liu, C.; Wu, Q.; Lu, J.; Liu, Y.; Shi, W. Selective Laser Melting-Wire Arc Additive Manufacturing Hybrid Fabrication of Ti-6Al-4V Alloy: Microstructure and Mechanical Properties. *Mater. Sci. Eng. A*, **2017**, *684*, 196–204. https://doi.org/10.1016/j.msea.2016.12.065.

[139] Zhang, L. C.; Klemm, D.; Eckert, J.; Hao, Y. L.; Sercombe, T. B. Manufacture by Selective Laser Melting and Mechanical Behavior of a Biomedical Ti–24Nb–4Zr–8Sn Alloy. *Scr. Mater.*, **2011**, *65* (1), 21–24. https://doi.org/10.1016/j.scriptamat.2011.03.024.

[140] Prakash, C.; Uddin, M. S. Surface Modification of β-Phase Ti Implant by Hydroxyapatite Mixed Electric Discharge Machining to Enhance the Corrosion Resistance and in-Vitro Bioactivity. *Surf. Coatings Technol.*, **2017**, *326*, 134–145. https://doi.org/10.1016/j.surfcoat.2017.07.040.

[141] Prakash, C.; Kansal, H. K.; Pabla, B. S.; Puri, S. Processing and Characterization of Novel Biomimetic Nanoporous Bioceramic Surface on β-Ti Implant by Powder Mixed Electric Discharge Machining. *J. Mater. Eng. Perform.*, **2015**, *24* (9), 3622–3633. https://doi.org/10.1007/s11665-015-1619-6.

[142] Prakash, C.; Kansal, H. K.; Pabla, B. S.; Puri, S. Experimental Investigations in Powder Mixed Electric Discharge Machining of Ti–35Nb–7Ta–5Zrβ-Titanium Alloy. *Mater. Manuf. Process.*, **2017**, *32* (3), 274–285. https://doi.org/10.1080/10426914.2016.1198018.

[143] Prakash, C.; Singh, S.; Pruncu, C.; Mishra, V.; Królczyk, G.; Pimenov, D.; Pramanik, A. Surface Modification of Ti-6Al-4V Alloy by Electrical Discharge Coating Process Using Partially Sintered Ti-Nb Electrode. *Materials*, **2019**, *12* (7), 1006. https://doi.org/10.3390/ma12071006.

[144] Pradhan, S.; Singh, S.; Prakash, C.; Królczyk, G.; Pramanik, A.; Pruncu, C. I. Investigation of Machining Characteristics of Hard-to-Machine Ti-6Al-4V-ELI Alloy for Biomedical Applications. *J. Mater. Res. Technol.*, **2019**, *8* (5), 4849–4862. https://doi.org/10.1016/j.jmrt.2019.08.033.

[145] Ahmed, T.; Rack, H. J. Phase Transformations during Cooling in A+β Titanium Alloys. *Mater. Sci. Eng. A*, **1998**, *243* (1–2), 206–211. https://doi.org/10.1016/S0921-5093(97)00802-2.

[146] Simonelli, M.; Tse, Y. Y.; Tuck, C. Effect of the Build Orientation on the Mechanical Properties and Fracture Modes of SLM Ti-6Al-4V. *Mater. Sci. Eng. A*, **2014**, *616*, 1–11. https://doi.org/10.1016/j.msea.2014.07.086.

[147] Xu, W.; Lui, E. W.; Pateras, A.; Qian, M.; Brandt, M. In Situ Tailoring Microstructure in Additively Manufactured Ti-6Al-4V for Superior Mechanical Performance. *Acta Mater.*, **2017**, *125*, 390–400. https://doi.org/10.1016/j.actamat.2016.12.027.

[148] Al-Bermani, S. S.; Blackmore, M. L.; Zhang, W.; Todd, I. The Origin of Microstructural Diversity, Texture, and Mechanical Properties in Electron Beam Melted Ti-6Al-4V. *Metall. Mater. Trans. A*, **2010**, *41* (13), 3422–3434. https://doi.org/10.1007/s11661-010-0397-x.

[149] Lin, C.-W.; Ju, C.-P.; Lin, J.-H. C. Comparison among Mechanical Properties of Investment-Cast c.p. Ti, Ti-6Al-7Nb and Ti-15Mo-1Bi Alloys. *Mater. Trans.*, **2004**, *45* (10), 3028–3032. https://doi.org/10.2320/matertrans.45.3028.

[150] Zhang, L.-C.; Attar, H.; Calin, M.; Eckert, J. Review on Manufacture by Selective Laser Melting and Properties of Titanium Based Materials for Biomedical Applications. *Mater. Technol.*, **2016**, *31* (2), 66–76. https://doi.org/10.1179/1753555715Y.0000000076.

[151] Lütjering, G. Influence of Processing on Microstructure and Mechanical Properties of (A+β) Titanium Alloys. *Mater. Sci. Eng. A*, **1998**, *243* (1–2), 32–45. https://doi.org/10.1016/S0921-5093(97)00778-8.

[152] Xu, W.; Sun, S.; Elambasseril, J.; Liu, Q.; Brandt, M.; Qian, M. Ti-6Al-4V Additively Manufactured by Selective Laser Melting with Superior Mechanical Properties. *JOM*, **2015**, *67* (3), 668–673. https://doi.org/10.1007/s11837-015-1297-8.

[153] Simonelli, M.; Tse, Y. Y.; Tuck, C. The Formation of α+β Microstructure in As-Fabricated Selective Laser Melting of Ti–6Al–4V. *J. Mater. Res.*, **2014**, *29* (17), 2028–2035. https://doi.org/10.1557/jmr.2014.166.

[154] Thijs, L.; Verhaeghe, F.; Craeghs, T.; Van Humbeeck, J.; Kruth, J.-P. A Study of the Microstructural Evolution during Selective Laser Melting of Ti–6Al–4V. *Acta Mater.*, **2010**, *58* (9), 3303–3312. https://doi.org/10.1016/j.actamat.2010.02.004.

[155] Murr, L. E.; Quinones, S. A.; Gaytan, S. M.; Lopez, M. I.; Rodela, A.; Martinez, E. Y.; Hernandez, D. H.; Martinez, E.; Medina, F.; Wicker, R. B. Microstructure and Mechanical Behavior of Ti–6Al–4V Produced by Rapid-Layer Manufacturing, for Biomedical Applications. *J. Mech. Behav. Biomed. Mater.*, **2009**, *2* (1), 20–32. https://doi.org/10.1016/j.jmbbm.2008.05.004.

[156] Sercombe, T.; Jones, N.; Day, R.; Kop, A. Heat Treatment of Ti-6Al-7Nb Components Produced by Selective Laser Melting. *Rapid Prototyp. J.*, **2008**, *14* (5), 300–304. https://doi.org/10.1108/13552540810907974.

[157] López, M..; Gutiérrez, A.; Jiménez, J. In Vitro Corrosion Behaviour of Titanium Alloys without Vanadium. *Electrochim. Acta*, **2002**, *47* (9), 1359–1364. https://doi.org/10.1016/S0013-4686(01)00860-X.

[158] Chlebus, E.; Kuźnicka, B.; Kurzynowski, T.; Dybała, B. Microstructure and Mechanical Behaviour of Ti–6Al–7Nb Alloy Produced by Selective Laser Melting. *Mater. Charact.*, **2011**, *62* (5), 488–495. https://doi.org/10.1016/j.matchar.2011.03.006.

[159] Aboulkhair, N. T.; Simonelli, M.; Parry, L.; Ashcroft, I.; Tuck, C.; Hague, R. 3D Printing of Aluminium Alloys: Additive Manufacturing of Aluminium Alloys Using Selective Laser Melting. *Prog. Mater. Sci.*, **2019**, *106*, 100578. https://doi.org/10.1016/j.pmatsci.2019.100578.

[160] Attar, H.; Bönisch, M.; Calin, M.; Zhang, L.-C.; Scudino, S.; Eckert, J. Selective Laser Melting of in Situ Titanium–Titanium Boride Composites: Processing, Microstructure and Mechanical Properties. *Acta Mater.*, **2014**, *76*, 13–22. https://doi.org/10.1016/j.actamat.2014.05.022.

[161] Gu, D.; Hagedorn, Y.-C.; Meiners, W.; Wissenbach, K.; Poprawe, R. Nanocrystalline TiC Reinforced Ti Matrix Bulk-Form Nanocomposites by Selective Laser Melting (SLM): Densification, Growth Mechanism and Wear Behavior. *Compos. Sci. Technol.*, **2011**, *71* (13), 1612–1620. https://doi.org/10.1016/j.compscitech.2011.07.010.

[162] Ma, Z..; Tjong, S..; Gen, L. In-Situ Ti-TiB Metal–Matrix Composite Prepared by a Reactive Pressing Process. *Scr. Mater.*, **2000**, *42* (4), 367–373. https://doi.org/10.1016/S1359-6462(99)00354-1.

[163] Morsi, K.; Patel, V. V. Processing and Properties of Titanium–Titanium Boride (TiBw) Matrix Composites—a Review. *J. Mater. Sci.*, **2007**, *42* (6), 2037–2047. https://doi.org/10.1007/s10853-006-0776-2.

[164] Attar, H.; Bönisch, M.; Calin, M.; Zhang, L. C.; Zhuravleva, K.; Funk, A.; Scudino, S.; Yang, C.; Eckert, J. Comparative Study of Microstructures and Mechanical Properties of in Situ Ti–TiB Composites Produced by Selective Laser Melting, Powder Metallurgy, and Casting Technologies. *J. Mater. Res.*, **2014**, *29* (17), 1941–1950. https://doi.org/10.1557/jmr.2014.122.

[165] Pattanayak, D. K.; Fukuda, A.; Matsushita, T.; Takemoto, M.; Fujibayashi, S.; Sasaki, K.; Nishida, N.; Nakamura, T.; Kokubo, T. Bioactive Ti Metal Analogous to Human Cancellous Bone: Fabrication by Selective Laser Melting and Chemical Treatments. *Acta Biomater.*, **2011**, *7* (3), 1398–1406. https://doi.org/10.1016/j.actbio.2010.09.034.

[166] Attar, H.; Prashanth, K. G.; Chaubey, A. K.; Calin, M.; Zhang, L. C.; Scudino, S.; Eckert, J. Comparison of Wear Properties of Commercially Pure Titanium Prepared by Selective Laser Melting and Casting Processes. *Mater. Lett.*, **2015**, *142*, 38–41. https://doi.org/10.1016/j.matlet.2014.11.156.

[167] Sun, J.; Yang, Y.; Wang, D. Mechanical Properties of a Ti6Al4V Porous Structure Produced by Selective Laser Melting. *Mater. Des.*, **2013**, *49*, 545–552. https://doi.org/10.1016/j.matdes.2013.01.038.

[168] Mordike, B..; Ebert, T. Magnesium: Properties — Applications — Potential. *Mater. Sci. Eng. A*, **2001**, *302* (1), 37–45. https://doi.org/10.1016/S0921-5093(00)01351-4.

[169] Somekawa, H.; Kinoshita, A.; Kato, A. Effect of Alloying Elements on Room Temperature Stretch Formability in Mg Alloys. *Mater. Sci. Eng. A*, **2018**, *732*, 21–28. https://doi.org/10.1016/j.msea.2018.06.098.

[170] Ding, Z.; Liu, W.; Sun, H.; Li, S.; Zhang, D.; Zhao, Y.; Lavernia, E. J.; Zhu, Y. Origins and Dissociation of Pyramidal $<c+A>$ Dislocations in Magnesium and Its Alloys. *Acta Mater.*, **2018**, *146*, 265–272. https://doi.org/10.1016/j.actamat.2017.12.049.

[171] Winzer, N.; Atrens, A.; Song, G.; Ghali, E.; Dietzel, W.; Kainer, K. U.; Hort, N.; Blawert, C. A Critical Review of the Stress Corrosion Cracking (SCC) of Magnesium Alloys. *Adv. Eng. Mater.*, **2005**, *7* (8), 659–693. https://doi.org/10.1002/adem.200500071.

[172] Jun, J. H.; Kim, J. M.; Park, B. K.; Kim, K. T.; Jung, W. J. Effects of Rare Earth Elements on Microstructure and High Temperature Mechanical Properties of ZC63 Alloy. *J. Mater. Sci.*, **2005**, *40* (9–10), 2659–2661. https://doi.org/10.1007/s10853-005-2099-0.

[173] Itoi, T.; Takahashi, K.; Moriyama, H.; Hirohashi, M. A High-Strength Mg–Ni–Y Alloy Sheet with a Long-Period Ordered Phase Prepared by Hot-Rolling. *Scr. Mater.*, **2008**, *59* (10), 1155–1158. https://doi.org/10.1016/j.scriptamat.2008.08.001.

[174] Johnston, S.; Shi, Z.; Atrens, A. The Influence of PH on the Corrosion Rate of High-Purity Mg, AZ91 and ZE41 in Bicarbonate Buffered Hanks' Solution. *Corros. Sci.*, **2015**, *101*, 182–192. https://doi.org/10.1016/j.corsci.2015.09.018.

[175] Toda-Caraballo, I.; Galindo-Nava, E. I.; Rivera-Díaz-del-Castillo, P. E. J. Understanding the Factors Influencing Yield Strength on Mg Alloys. *Acta Mater.*, **2014**, *75*, 287–296. https://doi.org/10.1016/j.actamat.2014.04.064.

[176] Agarwal, S.; Curtin, J.; Duffy, B.; Jaiswal, S. Biodegradable Magnesium Alloys for Orthopaedic Applications: A Review on Corrosion, Biocompatibility and Surface Modifications. *Mater. Sci. Eng. C*, **2016**, *68*, 948–963. https://doi.org/10.1016/j.msec.2016.06.020.

[177] Němec, M.; Jäger, A.; Tesař, K.; Gärtnerová, V. Influence of Alloying Element Zn on the Microstructural, Mechanical and Corrosion Properties of Binary Mg-Zn Alloys after Severe Plastic Deformation. *Mater. Charact.*, **2017**, *134*, 69–75. https://doi.org/10.1016/j.matchar.2017.10.017.

[178] Zhao, C.; Chen, X.; Pan, F.; Wang, J.; Gao, S.; Tu, T.; Liu, C.; Yao, J.; Atrens, A. Strain Hardening of As-Extruded Mg-XZn (x = 1, 2, 3 and 4 Wt%) Alloys. *J. Mater. Sci. Technol.*, **2019**, *35* (1), 142–150. https://doi.org/10.1016/j.jmst.2018.09.015.

[179] Jiang, J.; Ni, S.; Yan, H.; Wu, Q.; Song, M. New Orientations between B′2 Phase and α Matrix in a Mg-Zn-Mn Alloy Processed by High Strain Rate Rolling. *J. Alloys Compd.*, **2018**, *750*, 465–470. https://doi.org/10.1016/j.jallcom.2018.04.005.

[180] Ying, T.; Zheng, M. Y.; Li, Z. T.; Qiao, X. G.; Xu, S. W. Thermal Conductivity of As-Cast and as-Extruded Binary Mg–Zn Alloys. *J. Alloys Compd.*, **2015**, *621*, 250–255. https://doi.org/10.1016/j.jallcom.2014.09.199.

[181] Yan, Y.; Cao, H.; Kang, Y.; Yu, K.; Xiao, T.; Luo, J.; Deng, Y.; Fang, H.; Xiong, H.; Dai, Y. Effects of Zn Concentration and Heat Treatment on the Microstructure, Mechanical Properties and Corrosion Behavior of as-Extruded Mg-Zn Alloys Produced by Powder Metallurgy. *J. Alloys Compd.*, **2017**, *693*, 1277–1289. https://doi.org/10.1016/j.jallcom.2016.10.017.

[182] Farzadfar, S. A.; Martin, É.; Sanjari, M.; Essadiqi, E.; Wells, M. A.; Yue, S. On the Deformation, Recrystallization and Texture of Hot-Rolled Mg–2.9Y and Mg–2.9Zn Solid Solution Alloys—A Comparative Study. *Mater. Sci. Eng. A*, **2012**, *534*, 209–219. https://doi.org/10.1016/j.msea.2011.11.061.

[183] Němec, M.; Gärtnerová, V.; Jäger, A. Influence of Severe Plastic Deformation on Intermetallic Particles in Mg-12wt.%Zn Alloy Investigated Using Transmission Electron Microscopy. *Mater. Charact.*, **2016**, *119*, 129–136. https://doi.org/10.1016/j.matchar.2016.07.016.

[184] Yuasa, M.; Miyazawa, N.; Hayashi, M.; Mabuchi, M.; Chino, Y. Effects of Group II Elements on the Cold Stretch Formability of Mg–Zn Alloys. *Acta Mater.*, **2015**, *83*, 294–303. https://doi.org/10.1016/j.actamat.2014.10.005.

[185] Gieseke, M.; Noelke, C.; Kaierle, S.; Wesling, V.; Haferkamp, H. Selective Laser Melting of Magnesium and Magnesium Alloys. In *Magnesium Technology 2013*; Springer International Publishing: Cham, 2013; pp. 65–68. https://doi.org/10.1007/978-3-319-48150-0_11.

[186] Song, B.; Dong, S.; Deng, S.; Liao, H.; Coddet, C. Microstructure and Tensile Properties of Iron Parts Fabricated by Selective Laser Melting. *Opt. Laser Technol.*, **2014**, *56*, 451–460. https://doi.org/10.1016/j.optlastec.2013.09.017.

[187] Hu, D.; Wang, Y.; Zhang, D.; Hao, L.; Jiang, J.; Li, Z.; Chen, Y. Experimental Investigation on Selective Laser Melting of Bulk Net-Shape Pure Magnesium. *Mater. Manuf. Process.*, **2015**, *30* (11), 1298–1304. https://doi.org/10.1080/10426914.2015.1025963.

[188] Williams, J. C.; Starke, E. A. Progress in Structural Materials for Aerospace Systems. *Acta Mater.*, **2003**, *51* (19), 5775–5799. https://doi.org/10.1016/j.actamat.2003.08.023.

[189] Heinz, A.; Haszler, A.; Keidel, C.; Moldenhauer, S.; Benedictus, R.; Miller, W. Recent Development in Aluminium Alloys for Aerospace Applications. *Mater. Sci. Eng. A*, **2000**, *280* (1), 102–107. https://doi.org/10.1016/S0921-5093(99)00674-7.

[190] Xu, W.; Luo, Y.; Zhang, W.; Fu, M. Comparative Study on Local and Global Mechanical Properties of Bobbin Tool and Conventional Friction Stir Welded 7085-T7452 Aluminum Thick Plate. *J. Mater. Sci. Technol.*, **2018**, *34* (1), 173–184. https://doi.org/10.1016/j.jmst.2017.05.015.

[191] Tjong, S. Microstructural and Mechanical Characteristics of in Situ Metal Matrix Composites. *Mater. Sci. Eng. R Reports*, **2000**, *29* (3–4), 49–113. https://doi.org/10.1016/S0927-796X(00)00024-3.

[192] Li, W.; Yang, K.; Yin, S.; Yang, X.; Xu, Y.; Lupoi, R. Solid-State Additive Manufacturing and Repairing by Cold Spraying: A Review. *J. Mater. Sci. Technol.*, **2018**, *34* (3), 440–457. https://doi.org/10.1016/j.jmst.2017.09.015.

[193] Mishra, R. S.; Ma, Z. Y. Friction Stir Welding and Processing. *Mater. Sci. Eng. R Reports*, **2005**, *50* (1–2), 1–78. https://doi.org/10.1016/j.mser.2005.07.001.

[194] Brandl, E.; Heckenberger, U.; Holzinger, V.; Buchbinder, D. Additive Manufactured AlSi10Mg Samples Using Selective Laser Melting (SLM): Microstructure, High Cycle Fatigue, and Fracture Behavior. *Mater. Des.*, **2012**, *34*, 159–169. https://doi.org/10.1016/j.matdes.2011.07.067.

[195] Nie, F.; Dong, H.; Chen, S.; Li, P.; Wang, L.; Zhao, Z.; Li, X.; Zhang, H. Microstructure and Mechanical Properties of Pulse MIG Welded 6061/A356 Aluminum Alloy Dissimilar Butt Joints. *J. Mater. Sci. Technol.*, **2018**, *34* (3), 551–560. https://doi.org/10.1016/j.jmst.2016.11.004.

[196] Louvis, E.; Fox, P.; Sutcliffe, C. J. Selective Laser Melting of Aluminium Components. *J. Mater. Process. Technol.*, **2011**, *211* (2), 275–284. https://doi.org/10.1016/j.jmatprotec.2010.09.019.

[197] Ng, C. C.; Savalani, M. M.; Man, H. C.; Gibson, I. Layer Manufacturing of Magnesium and Its Alloy Structures for Future Applications. *Virtual Phys. Prototyp.*, **2010**, *5* (1), 13–19. https://doi.org/10.1080/17452751003718629.

[198] Ng, C. C.; Savalani, M. M.; Lau, M. L.; Man, H. C. Microstructure and Mechanical Properties of Selective Laser Melted Magnesium. *Appl. Surf. Sci.*, **2011**, *257* (17), 7447–7454. https://doi.org/10.1016/j.apsusc.2011.03.004.

[199] Savalani, M. M.; Pizarro, J. M. Effect of Preheat and Layer Thickness on Selective Laser Melting (SLM) of Magnesium. *Rapid Prototyp. J.*, **2016**, *22* (1), 115–122. https://doi.org/10.1108/RPJ-07-2013-0076.

[200] Zhang, B.; Liao, H.; Coddet, C. Effects of Processing Parameters on Properties of Selective Laser Melting Mg–9%Al Powder Mixture. *Mater. Des.*, **2012**, *34*, 753–758. https://doi.org/10.1016/j.matdes.2011.06.061.

[201] Manakari, V.; Parande, G.; Gupta, M. Selective Laser Melting of Magnesium and Magnesium Alloy Powders: A Review. *Metals*, **2016**, *7* (1), 2. https://doi.org/10.3390/met7010002.

[202] Das, S. Physical Aspects of Process Control in Selective Laser Sintering of Metals. *Adv. Eng. Mater.*, **2003**, *5* (10), 701–711. https://doi.org/10.1002/adem.200310099.

[203] Olakanmi, E. O. Selective laser sintering/melting (SLS/SLM) of Pure Al, Al–Mg, and Al–Si Powders: Effect of Processing Conditions and Powder Properties. *J. Mater. Process. Technol.*, **2013**, *213* (8), 1387–1405. https://doi.org/10.1016/j.jmatprotec.2013.03.009.

[204] Olakanmi, E. O.; Cochrane, R. F.; Dalgarno, K. W. A Review on Selective Laser Sintering/Melting (SLS/SLM) of Aluminium Alloy Powders: Processing, Microstructure, and Properties. *Prog. Mater. Sci.*, **2015**, *74*, 401–477. https://doi.org/10.1016/j.pmatsci.2015.03.002.

[205] Khan, M.; Dickens, P. Selective Laser Melting (SLM) of Gold (Au). *Rapid Prototyp. J.*, **2012**, *18* (1), 81–94. https://doi.org/10.1108/13552541211193520.

[206] Ion, J. *Laser Processing of Engineering Materials: Principles, Procedure and Industrial Application*; Elsevier: Amsterdam, The Netherlands, 2005.

[207] Singh, J.; Singh, C. Computational Analysis of Convective Heat Transfer across a Vertical Tube. *FME Trans.*, **2021**, *49*, 932–940. https://doi.org/10.5937/fme2104932S.

[208] Singh, J.; Singh, C. Numerical Analysis of Heat Dissipation from a Heated Vertical Cylinder by Natural Convection. *Proc. Inst. Mech. Eng. Part E J. Process Mech. Eng.*, **2017**, *231* (3), 405–413. https://doi.org/10.1177/0954408915600109.

[209] Singh, J.; Singh, C. Study of Buoyant Force Acting on Different Fluids Moving over a Horizontal Plate Due to Forced Convection. *IOP Conf. Ser. Mater. Sci. Eng.*, **2019**, *710* (1), 012044. https://doi.org/10.1088/1757-899X/710/1/012044.

[210] Khun, N. W.; Tan, A. W. Y.; Sun, W.; Liu, E. Effects of Nd:YAG Laser Surface Treatment on Tribological Properties of Cold-Sprayed Ti-6Al-4V Coatings Tested against 100Cr6 Steel under Dry Condition. *Tribol. Trans.*, **2019**, *62* (3), 391–402. https://doi.org/10.1080/10402004.2018.1563258.

[211] Astarita, A.; Genna, S.; Leone, C.; Minutolo, F. M. C.; Rubino, F.; Squillace, A. Study of the Laser Remelting of a Cold Sprayed Titanium Layer. *Procedia CIRP*, **2015**, *33*, 452–457. https://doi.org/10.1016/j.procir.2015.06.101.

[212] Rubino, F.; Astarita, A.; Carlone, P.; Genna, S.; Leone, C.; Memola Capece Minutolo, F.; Squillace, A. Selective Laser Post-Treatment on Titanium Cold Spray Coatings. *Mater. Manuf. Process.*, **2016**, *31* (11), 1500–1506. https://doi.org/10.1080/10426914.2015.1037912.

[213] Marrocco, T.; Hussain, T.; McCartney, D. G.; Shipway, P. H. Corrosion Performance of Laser Posttreated Cold Sprayed Titanium Coatings. *J. Therm. Spray Technol.*, **2011**, *20* (4), 909–917. https://doi.org/10.1007/s11666-011-9637-x.

[214] Poza, P.; Múnez, C. J.; Garrido-Maneiro, M. A.; Vezzù, S.; Rech, S.; Trentin, A. Mechanical Properties of Inconel 625 Cold-Sprayed Coatings after Laser Remelting. Depth Sensing Indentation Analysis. *Surf. Coatings Technol.*, **2014**, *243*, 51–57. https://doi.org/10.1016/j.surfcoat.2012.03.018.

[215] Kang, N.; Verdy, C.; Coddet, P.; Xie, Y.; Fu, Y.; Liao, H.; Coddet, C. Effects of Laser Remelting Process on the Microstructure, Roughness and Microhardness of in-Situ Cold Sprayed Hypoeutectic Al-Si Coating. *Surf. Coatings Technol.*, **2017**, *318*, 355–359. https://doi.org/10.1016/j.surfcoat.2017.01.057.

[216] Kim, D.; Chang, J.; Park, J.; Pak, J. J. Formation and Behavior of Kirkendall Voids within Intermetallic Layers of Solder Joints. *J. Mater. Sci. Mater. Electron.*, **2011**, *22* (7), 703–716. https://doi.org/10.1007/s10854-011-0357-2.

8 Sustainability Issues in Laser-Based Additive Manufacturing

Jashanpreet Singh
Chandigarh University

CONTENTS

8.1 ADDITIVE MANUFACTURING: BASIC INTRODUCTION

Currently, there is a lot of interest in "additive manufacturing," i.e., AM based on layer-by-layer building because of its potential to enable flexible manufacturing for end customers without the need for numerous supply chains. Producing complicated items with a short turnaround time and little to no waste thanks to AM is only one of the many benefits of this technique over the more common subtractive manufacturing [1–6]. Lightweight hollow items and moulds with internal cooling channels are two examples of AM's ability to maximize raw material savings without sacrificing quality. Carbon and greenhouse gas emissions may be reduced as a result of the end-use production process that eliminates several chains in favour of printed lightweight structures, which are particularly beneficial for the manufacturing of aeroplanes and vehicles [7–13]. AM technology is also used for the development and fabrication of complex parts, surface modification, and in joining and welding processes. Thermal spray [14–33], laser cladding [33–40], and friction stir processing [25,37,39,41–44] are adopted for the modification of the surfaces, which may be superior in terms of microstructural, mechanical, corrosion-resistant, and tribological properties [45–77,37–40]. AM technology is used for the post-processing of thermal spray, laser cladding, and friction stir processing joints. Complex-shaped components made up of metals, ceramics, and high-temperature materials may be treated using AM-based technology [12,37,39,57,58,66,67, 71–73,75,78–83]. The AM technology is also a big asset for the biomedical applications [84–88]. AM technology is widely used for the development of geopolymer catalyst [89], which is used for the synthesis of biofuels and crude oils [90–93].

Although AM has received a lot of attention as a promising alternative technology and has been used in a number of different fields, it still has room for improvement in terms of the types of materials it can print on, the accuracy of its prints, and the volume of products it can produce per print run [84,94,95]. General steps in AM include processing of the digital data, 3D modelling, 3D object creation, and post-processing [96]. 3D item creation procedures, such as vat photo-polymerization, material extrusion, powder bed fusion, and directed energy deposition, often dominate the performances of the final products [97]. At the precise location where the material to be melted, softened, or cured will be deposited, the energy must be delivered effectively. Then, the component is built through a technique of repeated layer deposition. As an example, during the extrusion process, the printing material is heated and then forced through a nozzle in order to reach its final destination. Low production rates are achieved by extruding molten material via a narrow nozzle, where energy transmission is sluggish owing to heat conduction over a tiny contact surface. Extrusion demands steady pressure of molten materials to maintain excellent printing resolution and high-quality surface finish. To avoid problems associated with deposition, energy may be transmitted directly to material that has already been positioned in the appropriate location. For this purpose, the laser is by far the most common energy source due to the fact that high-intensity laser beams projected onto the printing material may be absorbed effectively without the need for a transfer medium. Instant cure by photo-chemical/thermal reaction [98] (e.g. sintering or melting) are two possible outcomes of the laser energy [99–101], as shown in Figure 8.1. In contrast to incoherent sources such as thermal lamps or light-emitting diodes (LEDs), the spatially coherent light emitted by lasers enables the beams to distribute without critical divergence or loss of power over long distances. It also enables the beams to be focused into small spots, which in turn enables increased precision and throughput in the construction of 3D parts [102].

With its ability to produce high-complexity parts at cheap cost, AM has seen meteoric development over the last 30 years. The worldwide AM industry is estimated to be over $4.2 billion, which was approximately $0.25 billion in the mid-1990s [14]. AM is a key component of the so-called Industry 4.0 revolution. Since 1990, the market has grown at a compound annual rate (CAGR) of 25.4% [103]. The industry is expected to reach a value of anywhere between US $ 12.1 billion and over $20 billion by the end of 2020 [104,105]. In addition, sales of lasers for AM have been on the rise, with excellent year-over-year growth rates of 50.7% in 2015 compared to 2014 and a projected increase of 41.1% in 2016 compared to 2017 [106].

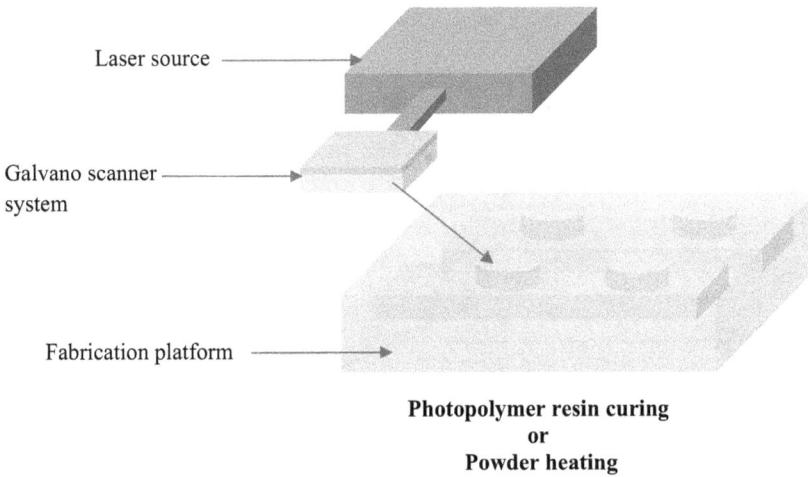

FIGURE 8.1 Schematic diagram of 3D printing machines for sintering or melting.

This high year-on-year growth that also represents the largest percentage growth especially in comparison with other laser categories is supporting the progressively increasing market share of laser technology in AM in broad market revenues (a rise of about 1% of total market revenue annually). This growth helps to explain why the market share is increasing at this rate. Growth in metal AM, which relies heavily on Yb-fibre lasers in two different processes, namely direct metal laser sintering (DMLS) and selective laser melting (SLM) machines, would be a major factor in driving the lasers in AM industry to new heights. By 2020, the metal AM market is projected to have grown from 2014's $0.16 billion to 2020's $0.78 billion (a CAGR of 31.5%) [107]. This chapter will go through the fundamentals of laser-based AM, the most important laser parameters that determine production performance, and several examples of laser-based AM techniques. We begin with a quick introduction to laser fundamentals, and then on to discuss the many kinds of lasers utilized in industrial 3D printing equipment, including CO_2 lasers, Nd:YAG lasers, Yb-fibre lasers, and excimer lasers.

8.2 LASERS IN AM TECHNOLOGY

Optical resonator, pumping power source, and gain medium are the three main components that make up a laser. The light beam is amplified by the gain medium within the optical resonator through stimulated emission utilizing pumping energy. Typically, lasers are categorized based on the kind of gain medium they use. Fibre lasers, solid-state lasers, and gas lasers are all examples of typical laser types utilized in AM, as illustrated in Figures 8.2–8.4. In the light of their prevalence in AM and other forms of precision manufacturing, CO_2, Nd:YAG, Yb-fibre, and excimer lasers are discussed as the typical lasers in this section.

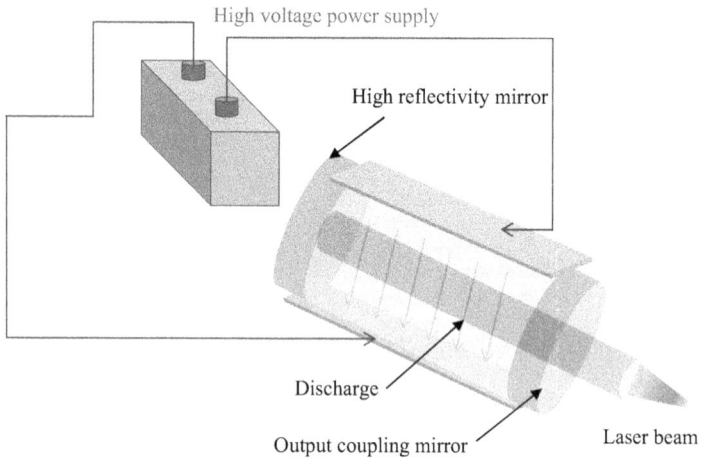

FIGURE 8.2   Schematic diagram of gas laser.

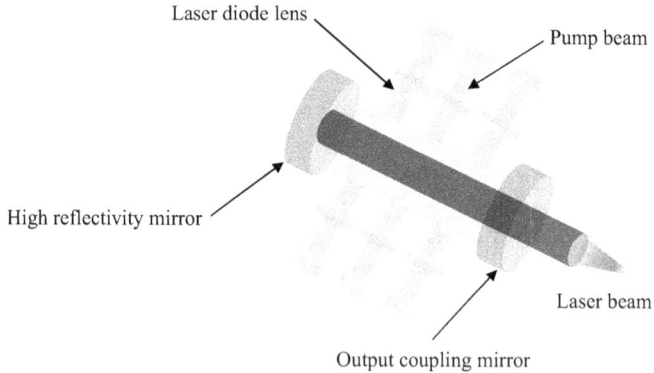

FIGURE 8.3   Schematic diagram of solid-state laser.

8.2.1   CARBON DIOXIDE (CO$_2$) LASER

Due to the fact that it was created in 1964 [109], the CO$_2$ laser is considered to be one of the earlier gas lasers. As can be seen in Figure 8.5, the laser is constructed using a discharge tube, an electric pump source, and a variety of optics, including mirrors, windows, and lenses, to construct an optical resonator. Additionally, the laser emits light in a coherent fashion. The gaseous gain medium in CO$_2$ lasers is carbon dioxide (CO$_2$), and it is pumped with a direct or alternating current (DC or AC) to cause population inversion and produce laser light [110]. CO$_2$ lasers have an infrared beam wavelength range of 9.0–11.0 m, with 10.6 m being the most popular choice for AM applications. Special materials, such as gold or silver for mirrors and Ge or ZnSe for windows and lenses, are utilized by the infrared wavelength emission [111].

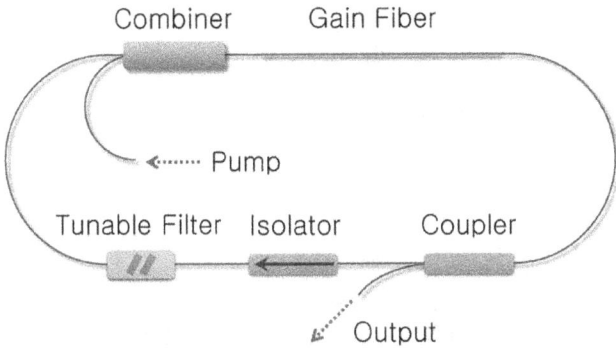

FIGURE 8.4 Schematic diagram of dual-wavelength rare-earth-doped fibre laser [108]. (Permissions under Creative Commons Attribution 4.0 International License.)

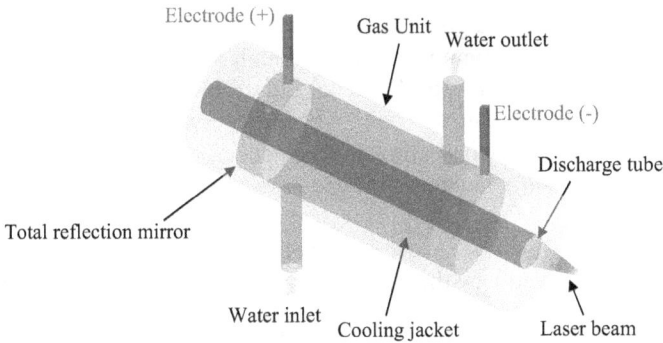

FIGURE 8.5 Schematic diagram of CO_2 laser.

CO_2 lasers are widely utilized in material processing, including but not limited to cutting, drilling, welding, and surface modification because of being highly efficient (5%–20%) and produce higher output power, i.e. 0.1–20 kW [112,113]. An electrically pumped gas discharge tube is sandwiched between a strong reflectivity mirror and a partially reflecting mirror (the "output coupler"). Using heat dissipation equipment like a cooling coil to cool the electrodes is essential when running at high power levels (often more than a few kilowatts). Cooling coils facilitate the conduction and convection to the laser system [114–116]. CO_2 lasers are beneficial in terms of precise production, cheap cost, great dependability, and their compactness made possible by the system's simplicity. The heat created by the process of pumping energy into a huge volume of CO_2 gas causes the laser structure to expand and contract, making the output power rather unstable. Turbulences in gas flow in the gas-assisted heat diffusion process may also contribute to the instability [117–119]. Entire optics should be evaluated for fatigue every 2000 hours in high power usage [120]. Limitations in functionality are caused by the long infrared operation wavelength. Due to its poor light absorptivity in the infrared range, the CO_2 laser has restricted throughput in metal component fabrication. Unfortunately, there are no optical fibres available for transporting this wavelength range; therefore, CO_2 lasers must rely on free-space bulk reflecting optics for beam output. Therefore, alternative kinds of lasers must be examined to deal with a broader range of materials.

8.2.2   Nd:YAG Solid-State Laser

These lasers are a kind of solid-state laser that use rod-shaped crystals of neodymium-doped yttrium aluminium garnet ($Nd_3+:Y_3Al_5O_{12}$). Gain medium consists of solid Nd:YAG crystals [121]. High-power Nd:YAG lasers are among the most widely utilized, alongside CO_2 lasers. Nd:YAG lasers, as illustrated in Figure 8.6 [122], generate a near-infrared (NIR) output wavelength of 1064 nm by optically pumping the gain medium in the radial direction with a flash lamp or axially with an 808 nm laser diode. One obvious benefit over the CO_2 laser is that the light beam may be carried by flexible optical fibres at this working wavelength, allowing for a smaller overall system footprint and greater delivery efficiency [119]. When the crystals are lightly doped with neodymium yttrium aluminium garnet, a Nd:YAG laser may be run in continuous as well as pulsed mode. Pulsed mode output power may reach up to 20 kW (pulse energy up to 120 J), whereas continuous mode output power is limited to a few kW [112]. Conventionally, the poor efficiency of electrical-to-optical power conversion is a problem for Nd:YAG lasers, which are often optically pumped by xenon flash lamps. Most of the unabsorbed energy is lost as heat, leading to poor beam quality [123]. This is due to the fact that thermal lensing and reflectivity effects are produced unexpectedly by the optical components when they are heated by thermal energy. Yet another flaw was the flash lamp's very low service life. Applying diode-pumped solid-state (DPSS) lasers may alleviate these drawbacks [124,125]. In comparison with lamp-pumped lasers, the total power efficiency of diode lasers may be increased by a factor of around five [126], thanks to the better electrical-to-optical

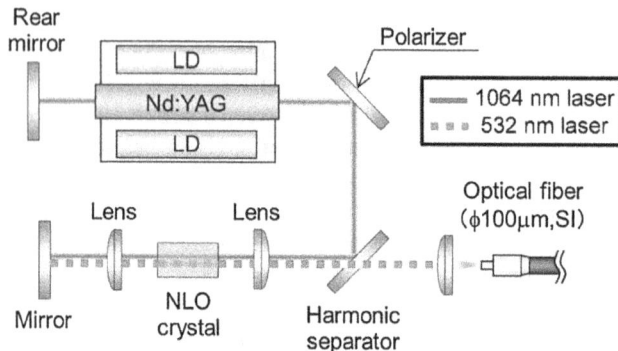

FIGURE 8.6   Schematic diagram of Nd:YAG laser [122]. (Permissions under Creative Commons Attribution 4.0 International License.)

power conversion efficiency of the laser diodes and selective stimulation of the gain medium. Smaller and more powerful Yb-fibre lasers are gradually replacing Nd:YAG lasers in AM. Notwithstanding, Nd:YAG lasers continue to see extensive application in research settings, particularly those concerned with parametric analysis [127–129] and the optimization of industrial parameters [130–132]. Nd:YVO₄ lasers have recently gained a lot of interest as a viable alternative to Nd:YAG lasers [133,134] due to their larger absorption band, lower operating threshold, and greater efficiency. Similar to how Nd:YAG lasers with a wavelength of 1064 nm are used to selectively cure photopolymer resins [135,136], Nd:YVO₄ lasers are used as a source of UV light through third harmonic generation (355 nm) in SLA47.

8.2.3 Yb-Doped Fibre Laser

Early on after their first invention, fibre lasers' pulse energy and output were subpar in comparison with bulk lasers. Due to rapid advancements in fibre lasers over the last decades, fibre lasers have emerged as the most favourable laser source as a substitute to conventional bulk lasers. Yb-fibres are best compared to other rare-earth-doped gain fibres due to the high power production and quantum efficiency (−94%) [137]. Yb-fibres are schematically illustrated in Figure 8.7 [138]. Because of their superior efficiency, Yb-fibre lasers have largely supplanted Nd:YAG lasers in AM [139]. The laser diodes in the 950–980 nm range serve as the pump, and the resulting 1030- to 1070-nm laser beams are in the NIR. The use of a fibre-based amplification medium and optical components results in a number of additional advantages, including a high ratio of electrical to optical efficiency (−25%), superior beam quality, resistance to environmental disturbances, and compactness of the system [112]. Since the laser light must go through the fibre, Yb-fibre lasers are similarly constrained in several ways. Light from bulk lasers travels through air, which is only somewhat effective as a light guide. When light is steered via an optical fibre, however, its nonlinear

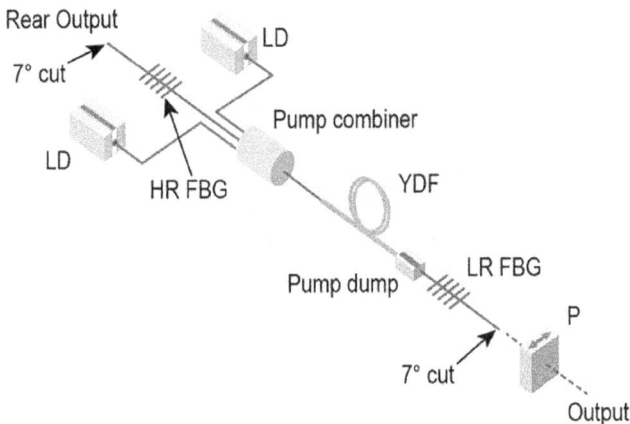

FIGURE 8.7 Schematic diagram of Yb-fibre laser [138]. (Permissions under Creative Commons Attribution 4.0 International License.)

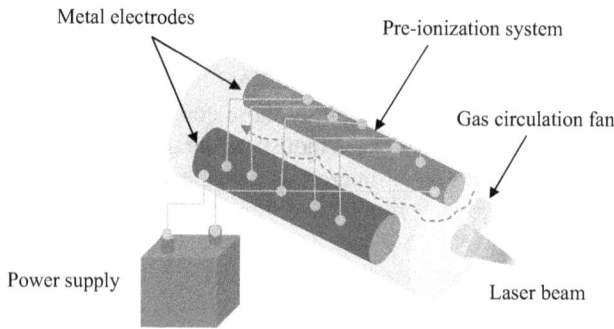

FIGURE 8.8 Schematic of diagram of an excimer laser.

characteristics are profoundly modified by the optical fibre itself. Self-focusing, the Raman effect, the Kerr lens effect, and self-phase modulation are all examples of optical nonlinear phenomena that might be hampered by the high peak power [140,141]. Alterations in polarization caused by bending, vibration, or temperature change in the fibre may also degrade laser power [142]. For applications demanding environmental stability, polarization-maintaining (PM) optical fibres are recommended.

8.2.4 Excimer Gas Laser

In order to generate nanosecond pulses in the ultraviolet (UV) range, excimer lasers use excimers as the gain medium and are powered by PED, i.e. pulsed electrical discharge. When referring to gas mixes, "excimer" is shorthand for "excited dimer," which refers to a combination of a noble gas (like Ar, Kr, or Xe), a halogen (like F or Cl), and a buffer gas (typically neon or helium). ArF, KrF, and XeCl lasers emitting beams having wavelengths of 193, 248, and 308 nm, respectively, are the most prominent excimer lasers in industrial applications [143]. These lasers have a broad operating wavelength range, from 157 to 351 nm (depending on the gas combination). As can be seen in Figure 8.8, much like other gas lasers (such as CO_2), an excimer laser requires a pump source, a gain medium, and an optical resonator to function. Only in the pulsed mode may excimer lasers be used; these lasers produce light bursts at a few kHz and 10–100 of watts. Most optical materials are UV-absorbent [144,145]; the production of UV-pulsed light is of major value in a wide range of industrial applications. Excimer lasers are not viable in AM machines because of their low beam quality, difficult maintenance, and highly high operation cost [146]. To generate laser light in the ultraviolet region, frequency-tripled Nd:YVO$_4$ lasers are the best option [147].

8.3 RESEARCH OPPORTUNITIES AND REQUIREMENT IN ENERGY AND SUSTAINABILITY

Ecological (or environmental) health, economic growth enabled by technological advancement, and social justice are at the heart of the concept of sustainability. Economic growth is often where engineering processes fall on the spectrum.

Material, energy consumption, usage of ecologically hazardous process enablers (such as cutting fluids and lubricants), and water use are the driving forces of economic growth in a manufacturing process [148]. To sum up, AM processes have favourable environmental properties [149]. As a result of using just what is necessary, AM technologies have the ability to minimize the lifecycle material mass and energy spent in comparison with traditional subtractive methods by reducing trash and the usage of detrimental auxiliary process enablers. Because they allow for the repair and remanufacture of outdated or broken tools, many AM methods also have the potential to eradicate supply chain procedures connected with the fabrication of new tooling [150]. Many different industries have discovered that by remanufacturing items instead of making new ones, they may save money, boost productivity, and lessen their negative effects on the environment. With the use of AM technology, the tooling sector has been able to remanufacture and repair high-value tools and dies without contributing to environmental pollution [151]. Billet is used to make a wide variety of CNC-machined items. The volume of the billet may be much larger than the finished product. This is the so-called buy-to-fly ratio in the aircraft industry. Materials used in AM techniques are very resource efficient, lowering the buy-to-fly ratio by a considerable amount [152] since they only consume the raw material needed to solidify the finished product and, in certain circumstances, a tiny quantity of support structure.

There are five main environmental and sustainability advantages of adopting AM [153], all of which contribute to a less carbon footprint. This means less raw materials will be needed elsewhere in the supply chain. This means less time and energy spent on primary material ore mining and processing. The demand for energy- and resource-intensive production techniques, including casting and CNC machining with cutting fluids, is low. The skill to create items with enhanced efficiency and functionality is required, such as hydraulic parts with conformal fluid pathways. Lessening the load on the vehicle due to lighter transportation items helps to reduce emissions. Producing components nearer to where they would be used would save shipping costs.

Different facets of AM's long-term viability are discussed here. The first part of this paper is a summary of the workshop's results on energy and sustainability from 2009. This paper discusses the prospects and requirements for future research in the field of sustainable AM. In the paper's second section, an analysis of the energy consumption of a 3D Systems Vanguard HiQ sinterstation is presented in detail. Results are reviewed in the light of comparisons with some other laser sintering/melting systems, and energy consumption is broken down by system component. The third portion of the paper explains how to use electrolytic deposition techniques to make metal and ceramic structural components. In this case, selective laser sintering is used to create a porous "green" component (SLS). Electrochemical deposition (or plating) of a metallic phase in an electrolytic solution removes the continuous internal porosity after appropriate low-temperature post-processing to transform or eliminate the binder. Energy costs may be drastically reduced since the process itself does not need a high temperature.

On March 30–31, 2009, the National Science Foundation hosted the Roadmap for AM (RAM) Workshop in the greater Washington, DC region. About 65 invited participants spoke about various aspects of AM and identified potential future research directions. Dr. Hong-Chao Zhang of Texas Tech University presided over the sustainability thematic breakout session. Many people contributed to the sustainability effort, including Bert Bras from Georgia Tech, I.S. Jawahir from the University of Kentucky, Frank Liou from Missouri University of Science and Technology, Rhett Mayor from Georgia Tech, Jyoti Mazumder from the University of Michigan-Ann Arbor, John Kawola from ZCorp, Greg Morris from Morris Technologies, and Phil Reeves from Morris Technologies (Econolyst Ltd.). This paper provides a brief overview of the "Energy and Sustainability" section of the workshop's final whole Roadmap Report [154]. When compared to more conventional production methods like casting, milling, and moulding, AM procedures have the potential to have a beneficial effect on environmental friendliness. In many circumstances, the rate of scrap materials produced by AM is so low that it does not even warrant mention. For certain applications, such as those that do not need prolonged periods of processing at high temperatures, AM may have a very little impact on energy consumption. For example, cutting fluids, casting release compounds, and forging lubricants, all of which have a negative effect on sustainability, are not necessary in AM. In particular, automated component repair is well suited to AM procedures based on metallic powder deposition techniques (e.g. LENS, LAMP, DMD). When a broken or worn-out component is fixed and put back into operation instead of being thrown away, a significant quantity of resources is conserved. When it comes to manufacturing, the whole tooling supply chain disappears. Due to the regional and decentralized nature of AM production, less energy and less greenhouse gas emissions are used since fewer vehicles need to be used. Last but not least, the flexibility of AM's design lets us make components that have better energy efficiency after they are put into operation. Included are examples such as conformal cooling channels, gas flow routes, streamlined shape, lightweight cellular pieces, etc. There has not been a lot of research on the whole life cycle of AM processes. The environmental impact of producing AM components should take into account more than just the resources used and the waste generated. The energy costs of the supply chain, the effects of product use, and issues of disposal and recycling at the end of a product's life are all crucial factors to consider. It is necessary to use support structures in certain AM processes in order to preserve geometry when creating parts. There has been no thought given to how discarding support materials could affect the natural world.

Sustainability studies that may be applied to AM processes can cover a lot of ground. Design and development efforts in the energy and aerospace sectors, in particular, may help accelerate the spread of AM goods into mainstream commerce. The performance of the whole "gate-to-grave" phase of the life cycle may be improved by using AM to make components with the highest possible sustainability standards. Particularly for metals and plastics, consolidating several components into a single AM item may ease recycling and disposal. It is a potential line of inquiry to investigate the possibility of producing AM feedstock from by-products of other manufacturing processes. It might include, for instance, using machined chips either immediately or after little processing. Materials for AM that are inherently safe for the planet need to be developed. They must be harmless to the environment and biodegradable. The AM process should be seen in the context of general sustainability issues. Lifecycle studies from "cradle to gate" and "cradle to grave" and the creation of environmentally friendly supply networks are among them. Predicting and evaluating AM product sustainability is another area of investigation. It would be beneficial to have a model for lifecycle value engineering. To maximize the benefits of AM, it is preferable to include principles of sustainability into design processes at the same time as regulations and standards for AM are being established. For a quantitative assessment of AM components' sustainability, we need fair indicators. Applying AM components in settings where comparative models between their energy use and environmental performance and those of traditionally made items are available would be beneficial.

8.4   SUSTAINABILITY PROBLEMS IN LASER-BASED AM

8.4.1   RESEARCH NEED IN ENERGY USE

Many variables, including feedstocks, process variations, load and usage patterns, and regional distributions, contribute to the vast range of energy efficiency in AM processes. Distributed production, made possible by AM, supposedly facilitates the use of renewable sources of energy [155]. There have been several energy studies on various AM technologies, including laser sintering and FDM. However, the majority of these studies have been conducted in academic contexts rather than commercial ones. The difficulty in comparing findings stems from the fact that experimental methods and measuring methodologies differ from study to study. Energy requirements for the production of AM feedstocks, such as powder, resin, and filaments, are often high. Energy content of these feedstocks is often assessed with considerable errors, which is problematic. As an added downside, energy efficiency improvements have mostly been made in niche areas, such as with specialized machinery.

The following are examples of areas where further study is needed: a unified approach to quantifying the power requirements of AM operations; a more accurate assessment of the inherent energy in a number of AM feedstocks; the effect that decentralized AM has on energy consumption; and recommendations for maximizing energy efficiency in printing and disseminating the results of such endeavours within AM communities.

8.4.2   RESEARCH NEED IN OCCUPATIONAL HEALTH

In particular, particulate matter and volatile organic compounds (VOCs) emitted by AM machines have been cited as potential hazards to worker health. However, there has been a lack of effort put into defining and quantifying these air pollutants. There is a growing possibility of encountering familiar risks in atypical contexts due to the usage of unconventional feedstocks and innovative procedures. Because AM has made mass production of goods more affordable for smaller businesses, the locations, people, and circumstances often connected with AM technology are beginning to diverge from those typically associated with more traditional manufacturing methods. Even in settings with less than ideal ventilation and emission control, such as classrooms and start-up prototype labs, makerspaces, basements, and garages, the usage of advanced AM methods is on the rise. Many of the health and safety issues here are probably not all that different from those seen in other industrial processes, but the institutional understanding and capability to deal with them are different. The social and economic advantages of AM depend on its safe and responsible use, but this relies on a new set of difficulties and uncertainties brought about by the combination of traditional hazards in unorthodox contexts, potential untraditional hazards, and overexposure by the risk-naive communities. To fully reap the advantages of AM technologies, it is crucial to understand and address both the potential hazards and advantages of AM to people across the board.

The following are the areas where further study is needed: awareness of the processes behind the aetiology of pollution; how feedstock ingredients, printing parameters, and machine settings are related to emissions; experience with exposure identification and monitoring, with a focus on indoor air quality, whether at school or at home; the development of emission controls and "safer-by-design" techniques for AM is a focus of current R&D efforts; creation of more secure social norms; risk analysis and administration is a focus; and a new approach to teaching workers about safety in the workplace.

8.4.3   Research Need in Waste

There is a lack of comprehensive data on the materials and amounts of waste produced by AM procedures and 3D-printed objects after they have reached the end of their useful lives. Research is required to quantify existing and future AM practices. It is tough to get into more precise topics without that data. Current research should focus on the materials used in AM, the circumstances under which they are utilized, and the rates at which waste is generated in various production contexts and at the end of their useful lives.

Because of this, there are a number of questions that require answering via study: the categorization of waste from production in commercial and noncommercial contexts; the creation of an international resin labelling system for filaments [156] to standardize the handling and cycling of construction materials; products' end-of-life administration in a globally dispersed production system; and creation and dissemination of guidelines for waste reduction in AM communities. The factors that promote and impede AM's ability to lengthen the lifespan of a product will be investigated. Examining whether AM's ability to create innovative forms, products, and components presents any special difficulties in recycling, remanufacturing, or end-of-life waste management.

8.4.4   Research Need in Lifecycle Assessment

Many assumptions and simplifications underpin the current state of LCA research on AM processes or AM products. There are often contrasting findings in the published literature. There are large information gaps in both the upstream (the production of feedstock) and the downstream (the disposal of used materials) stages of the supply chain. Furthermore, much of the information gathered comes from lab situations rather than real-world ones. Each of these factors reduces the impact of LCA studies on enhancing products and processes or developing new policies. Among the areas where more study is needed are:

- Standardized evaluation of AM and traditional manufacturing's energy and material usage.
- Different AM routes' supply chain environmental footprints may be established with the use of certain methodologies that include data collecting and exchange.
- Potential outcomes for the natural world as a result of developments in the supply chain made possible by AM.
- The results of widespread use of AM equipment.
- For a full evaluation of AM's impact during its lifetime, it is important to collect and share relevant data.
- After the widespread use of AM technology, retirement plans may need to be revised.

8.4.5   Research Needs in Cross-Cutting and Policy

Because of its decentralized character and close relationship to product design, AM has several distinctive properties with respect to ecological footprints. To fully achieve AM's potential benefits to sustainability and reduce the environmental consequences, it may need unique methodologies and cross-cutting research across issue areas including integrated product and process development [157] and standardized characterization [158]. To fully grasp the implications of AM on resource use, we need to investigate the production systems and value chain that will be built on this technology [159]. To further understand how AM might allow more sustainable ways of production and consumption, six study topics have been developed from the United Kingdom's viewpoint [160]: design, supply chains, information flows, entrepreneurship, business models, and education. Also, similar to other new technologies, AM has the potential to cause a wide range of policy concerns, including those related to public safety, product quality, and the well-being of AM workers. Studies of lifecycles, health, energy, and/or waste may provide answers to certain policy and contextual problems, but there are others that will need more investigation. The following gaps in knowledge have been identified:

- Increased consideration of environmental impact when developing AM products.
- Multidisciplinary AM processes and distributed manufacturing policy concerns.
- Coding and standards for AM feedstock.
- Examining how people view AM's many applications.

8.5   CONCLUDING REMARKS AND FUTURE RESEARCH DIRECTIONS

Selecting and using industrial processes with a focus on sustainability has become more important. Although few AM activities have been explicitly examined for environmental and energy effects, the same holds true for AM processes. The CO_2 and Nd:YAG lasers have formed the backbone of the laser manufacturing industry, powering not just AM but also other laser-based production methods. Increases in average power, system stability, parametric tunability at a high level, and cheap maintenance costs have made Yb-fibre lasers a viable alternative to Nd:YAG lasers. Excimer lasers may be utilized in AM that calls for high-power UV laser beams despite the fact that they have relatively poor beam quality and are more expensive. Light-matter interaction with varying operating wavelengths, light strength, pulse duration, and beam quality was taken into account to comprehend laser-based AM's manufacturing capabilities, including printable material, accuracy, and throughput. It is important to choose the AM laser source based on the desired performance outcomes. Laser technologies will continue to play an important role in the future of AM. The expectation is that laser-based AM will continue to displace subtractive manufacturing methods, augment existing methods to boost their efficiency (a process known as "hybrid manufacturing"), or give rise to whole new economic sectors. There was also an effort to develop a uniform method for evaluating AM's energy use in production. Using future-oriented, credible AM scenario planning is essential for this study. Toxicology of emissions, exposure control strategies, and exposure assessment are all areas where further study is needed to better understand and manage AM's risks to workers' health. One possible outcome of this effort is the creation of "safer-by-design" guidelines, resources, and methods for use in AM. More study is required into novel materials, quantification of the energy implications of AM's end products, and the creation of standardized procedures to compare the energy use of additive with those of traditional manufacturing processes. Lightweight materials, batteries, insulation, and energy generation technologies might all be on the table for this effort if it involves AM methods. Studies examining the potential for novel forms, products, and components made feasible by AM to provide new issues for waste management, as well as studies examining industrial production waste, are needed. Issues that go across disciplines and policies were also recognized, as were the need for further investigation into the potential legal and ethical ramifications of bioprinting. Other aspects include the application of lifecyle analysis by diverse component groups, the public's perception of different AM uses, and the early identification of dangers peculiar to AM in "desktop" contexts. Notable efforts and events have also been organized by the research and development community to address environmental implications of AM, particularly in the area of occupational health and safety science as it pertains to the many different AM processes, such as fused deposition modelling (FDM) and powder bed fusion.

REFERENCES

1. Yoon, H.-S., Lee, J.-Y., Kim, H.-S., Kim, M.-S., Kim, E.-S., Shin, Y.-J., Chu, W.-S., Ahn, S.-H.: A comparison of energy consumption in bulk forming, subtractive, and additive processes: Review and case study. *Int. J. Precis. Eng. Manuf. Technol.* 1, 261–279 (2014). https://doi.org/10.1007/s40684-014-0033-0.
2. Ahn, S.-H., Chun, D.-M., Chu, W.-S.: Perspective to green manufacturing and applications. *Int. J. Precis. Eng. Manuf.* 14, 873–874 (2013). https://doi.org/10.1007/s12541-013-0114-y.
3. Moon, S.K., Tan, Y.E., Hwang, J., Yoon, Y.-J.: Application of 3D printing technology for designing light-weight unmanned aerial vehicle wing structures. *Int. J. Precis. Eng. Manuf. Technol.* 1, 223–228 (2014). https://doi.org/10.1007/s40684-014-0028-x.
4. Ko, H., Moon, S.K., Hwang, J.: Design for additive manufacturing in customized products. *Int. J. Precis. Eng. Manuf.* 16, 2369–2375 (2015). https://doi.org/10.1007/s12541-015-0305-9.
5. Singh, J., Singh, S.: A review on Machine learning aspect in physics and mechanics of glasses. *Mater. Sci. Eng. B.* 284, 115858 (2022). https://doi.org/10.1016/j.mseb.2022.115858.
6. Singh, R., Toseef, M., Kumar, J., Singh, J.: Benefits and challenges in additive manufacturing and its applications. In: Kaushal, S., Singh, I., Singh, S., and Gupta, A. (eds.) *Sustainable Advanced Manufacturing and Materials Processing.* pp. 137–157. CRC Press, Boca Raton (2022). https://doi.org/10.1201/9781003269298-8.
7. Khare, V., Ruby, C., Sonkaria, S., Taubert, A.: A green and sustainable nanotechnology: Role of ionic liquids. *Int. J. Precis. Eng. Manuf.* 13, 1207–1213 (2012). https://doi.org/10.1007/s12541-012-0160-x.
8. Shan, Z., Qin, S., Liu, Q., Liu, F.: Key manufacturing technology & equipment for energy saving and emissions reduction in mechanical equipment industry. *Int. J. Precis. Eng. Manuf.* 13, 1095–1100 (2012). https://doi.org/10.1007/s12541-012-0143-y.
9. Ahn, S.-H.: An evaluation of green manufacturing technologies based on research databases. *Int. J. Precis. Eng. Manuf. Technol.* 1, 5–9 (2014). https://doi.org/10.1007/s40684-014-0001-8.
10. Lee, G., Sul, S.-K., Kim, J.: Energy-saving method of parallel mechanism by redundant actuation. *Int. J. Precis. Eng. Manuf. Technol.* 2, 345–351 (2015). https://doi.org/10.1007/s40684-015-0042-7.
11. Huang, S.H., Liu, P., Mokasdar, A., Hou, L.: Additive manufacturing and its societal impact: A literature review. *Int. J. Adv. Manuf. Technol.* 67, 1191–1203 (2013). https://doi.org/10.1007/s00170-012-4558-5.
12. Prashar, G., Vasudev, H.: A comprehensive review on sustainable cold spray additive manufacturing: State of the art, challenges and future challenges. *J. Clean. Prod.* 310, 127606 (2021). https://doi.org/10.1016/j.jclepro.2021.127606.
13. Prashar, G., Vasudev, H., Bhuddhi, D.: Additive manufacturing: Expanding 3D printing horizon in industry 4.0. *Int. J. Interact. Des. Manuf.* (2022). https://doi.org/10.1007/s12008-022-00956-4.
14. Singh, J., Kumar, S., Mohapatra, S.K.: Tribological analysis of WC–10Co–4Cr and Ni–20Cr$_2$O$_3$ coating on stainless steel 304. *Wear.* 376–377, 1105–1111 (2017). https://doi.org/10.1016/j.wear.2017.01.032.
15. Kumar, K., Kumar, S., Singh, G., Singh, J., Singh, J.: Erosion wear investigation of HVOF sprayed WC-10Co4Cr coating on slurry pipeline materials. *Coatings.* 7, 54 (2017). https://doi.org/10.3390/coatings7040054.
16. Singh, J., Singh, S., Pal Singh, J.: Investigation on wall thickness reduction of hydropower pipeline underwent to erosion-corrosion process. *Eng. Fail. Anal.* 127, 105504 (2021). https://doi.org/10.1016/j.engfailanal.2021.105504.

17. Singh, J., Singh, S.: Neural network supported study on erosive wear performance analysis of Y_2O_3/WC-10Co4Cr HVOF coating. *J. King Saud Univ. - Eng. Sci.* (2022). https://doi.org/10.1016/j.jksues.2021.12.005.

18. Singh, J.: Application of thermal spray coatings for protection against erosion, abrasion, and corrosion in hydropower plants and offshore industry. In: Thakur, L. and Vasudev, H. (eds.) *Thermal Spray Coatings.* pp. 243–283. CRC Press, Boca Raton (2021). https://doi.org/10.1201/9781003213185-10.

19. Singh, J., Singh, J.P.: Performance analysis of erosion resistant Mo_2C reinforced WC-CoCr coating for pump impeller with Taguchi's method. *Ind. Lubr. Tribol.* 74, 431–441 (2022). https://doi.org/10.1108/ILT-05-2020-0155.

20. Singh, J.: Slurry erosion performance analysis and characterization of high-velocity oxy-fuel sprayed Ni and Co hardsurfacing alloy coatings. *J. King Saud Univ. - Eng. Sci.* (2021). https://doi.org/10.1016/j.jksues.2021.06.009.

21. Singh, S., Garg, J., Singh, P., Singh, G., Kumar, K.: Effect of hard faced Cr-alloy on abrasive wear of low carbon rotavator blades using design of experiments. *Mater. Today Proc.* 5, 3390–3395 (2018). https://doi.org/10.1016/j.matpr.2017.11.583.

22. Singh, J., Kumar, S., Mohapatra, S.K.: Optimization of erosion wear influencing parameters of HVOF sprayed pumping material for coal-water slurry. *Mater. Today Proc.* 5, 23789–23795 (2018). https://doi.org/10.1016/j.matpr.2018.10.170.

23. Singh, J., Kumar, S., Singh, G.: Taguchi's approach for optimization of tribo-resistance parameters Forss304. *Mater. Today Proc.* 5, 5031–5038 (2018). https://doi.org/10.1016/j.matpr.2017.12.081.

24. Singh, J.: *Investigation on Slurry Erosion of Different Pumping Materials and Coatings.* Thapar Institute of Engineering and Technology, Patiala, India (2019).

25. Sharma, Y., Singh, K.J., Vasudev, H.: Experimental studies on friction stir welding of aluminium alloys. *Mater. Today Proc.* 50, 2387–2391 (2022). https://doi.org/10.1016/j.matpr.2021.10.254.

26. Singh, J.: Analysis on suitability of HVOF sprayed Ni-20Al, Ni-20Cr and Al-20Ti coatings in coal-ash slurry conditions using artificial neural network model. *Ind. Lubr. Tribol.* 71, 972–982 (2019). https://doi.org/10.1108/ILT-12-2018-0460.

27. Singh, J., Kumar, S., Mohapatra, S.K.: Tribological performance of Yttrium (III) and Zirconium (IV) ceramics reinforced WC–10Co4Cr cermet powder HVOF thermally sprayed on X2CrNiMo-17-12-2 steel. *Ceram. Int.* 45, 23126–23142 (2019). https://doi.org/10.1016/j.ceramint.2019.08.007.

28. Singh, J., Kumar, S., Mohapatra, S.K.: Erosion wear performance of Ni-Cr-O and NiCrBSiFe-WC(Co) composite coatings deposited by HVOF technique. *Ind. Lubr. Tribol.* 71, 610–619 (2019). https://doi.org/10.1108/ILT-04-2018-0149.

29. Singh, J., Kumar, S., Mohapatra, S.K.: Erosion tribo-performance of HVOF deposited Stellite-6 and Colmonoy-88 micron layers on SS-316L. *Tribol. Int.* 147, 105262 (2020). https://doi.org/10.1016/j.triboint.2018.06.004.

30. Singh, J., Kumar, S., Mohapatra, S.K.: An erosion and corrosion study on thermally sprayed WC-Co-Cr powder synergized with Mo_2C/Y_2O_3/ZrO_2 feedstock powders. *Wear.* 438–439, 102751 (2019). https://doi.org/10.1016/j.wear.2019.01.082.

31. Singh, J., Singh, S.: Neural network prediction of slurry erosion of heavy-duty pump impeller/casing materials 18Cr-8Ni, 16Cr-10Ni-2Mo, super duplex 24Cr-6Ni-3Mo-N, and grey cast iron. *Wear.* 476, 203741 (2021). https://doi.org/10.1016/j.wear.2021.203741.

32. Singh, J.: Wear performance analysis and characterization of HVOF deposited Ni-20Cr_2O_3, Ni-30Al_2O_3, and Al_2O_3–13TiO_2 coatings. *Appl. Surf. Sci. Adv.* 6, 100161 (2021). https://doi.org/10.1016/j.apsadv.2021.100161.

33. Singh, J.: Tribo-performance analysis of HVOF sprayed 86WC-10Co4Cr & Ni-Cr_2O_3 on AISI 316L steel using DOE-ANN methodology. *Ind. Lubr. Tribol.* 73, 727–735 (2021). https://doi.org/10.1108/ILT-04-2020-0147.

34. Singh, J., Singh, S., Gill, R.: Applications of biopolymer coatings in biomedical engineering. *J. Electrochem. Sci. Eng.* (2022). https://doi.org/10.5599/jese.1460.
35. Vasudev, H., Singh, P., Thakur, L., Bansal, A.: Mechanical and microstructural characterization of microwave post processed Alloy-718 coating. *Mater. Res. Express.* 6, 1265f5 (2020). https://doi.org/10.1088/2053-1591/ab66fb.
36. Vasudev, H., Thakur, L., Singh, H., Bansal, A.: Erosion behaviour of HVOF sprayed Alloy718-nano Al_2O_3 composite coatings on grey cast iron at elevated temperature conditions. *Surf. Topogr. Metrol. Prop.* 9, 035022 (2021). https://doi.org/10.1088/2051-672X/ac1c80.
37. Prashar, G., Vasudev, H., Thakur, L.: Performance of different coating materials against slurry erosion failure in hydrodynamic turbines: A review. *Eng. Fail. Anal.* 115, 104622 (2020). https://doi.org/10.1016/j.engfailanal.2020.104622.
38. Vasudev, H., Prashar, G., Thakur, L., Bansal, A.: Electrochemical corrosion behavior and microstructural characterization of HVOF sprayed Inconel-718 coating on gray cast iron. *J. Fail. Anal. Prev.* 21, 250–260 (2021). https://doi.org/10.1007/s11668-020-01057-8.
39. Prashar, G., Vasudev, H., Thakur, L.: Influence of heat treatment on surface properties of HVOF deposited WC and Ni-based powder coatings: A review. *Surf. Topogr. Metrol. Prop.* 9, 043002 (2021). https://doi.org/10.1088/2051-672X/ac3a52.
40. Singh, G., Vasudev, H., Bansal, A., Vardhan, S., Sharma, S.: Microwave cladding of Inconel-625 on mild steel substrate for corrosion protection. *Mater. Res. Express.* 7, 026512 (2020). https://doi.org/10.1088/2053-1591/ab6fa3.
41. Wang, Y., Stella, J., Darut, G., Poirier, T., Liao, H.: APS prepared NiCrBSi-YSZ composite coatings for protection against cavitation erosion. *J. Alloys Compd.* 699, 1095–1103 (2017). https://doi.org/10.1016/j.jallcom.2017.01.034.
42. Chauhan, S.R., Gokul Krishna, U., Setia, S.: Finite element based simulation model for micro turning of nanoparticle-reinforced aluminum alloy (7075-T6) composite. In: *Manufacturing Engineering: Select Proceedings of CPIE 2019.* pp. 561–574. Springer, Singapore (2020). https://doi.org/10.1007/978-981-15-4619-8_40.
43. Singh, G.: A review on different high velocity oxyfuel coated matrix materials. *Mater. Today Proc.* 37, 2294–2297 (2020). https://doi.org/10.1016/j.matpr.2020.07.726.
44. Gloria, A., Montanari, R., Richetta, M., Varone, A.: Alloys for aeronautic applications: State of the art and perspectives. *Metals.* 9, 1–26 (2019). https://doi.org/10.3390/met9060662.
45. Kumar, S., Singh, M., Singh, J., Singh, J.P., Kumar, S.: Rheological characteristics of Uni/Bi-variant particulate iron ore slurry: Artificial neural network approach. *J. Min. Sci.* 55, 201–212 (2019). https://doi.org/10.1134/S1062739119025468.
46. Singh, J., Singh, R., Singh, S., Vasudev, H., Kumar, S.: Reducing scrap due to missed operations and machining defects in 90PS pistons. *Int. J. Interact. Des. Manuf.* (2022). https://doi.org/10.1007/s12008-022-01071-0.
47. Singh, J., Singh, J.P., Singh, M., Szala, M.: Computational analysis of solid particle-erosion produced by bottom ash slurry in 90° elbow. In: *MATEC Web of Conferences.* Vol. 252, p. 04008 (2019). https://doi.org/0.1051/matecconf/201925204008.
48. Kumar, S., Singh, J., Mohapatra, S.K.: Role of particle size in assessment of physico-chemical properties and trace elements of Indian fly ash. *Waste Manag. Res.* 36, 1016–1022 (2018). https://doi.org/10.1177/0734242X18804033.
49. Kumar, S., Singh, J., Mohapatra, S.K.: Influence of particle size on leaching characteristic of fly ash. In: *15th International Conference on Environmental Science and Technology, 31 August to 2 September 2017.* p. 01243. Rhodes, Greece (2017).
50. Singh, J., Kumar, S., Mohapatra, S.K.: Study on solid particle erosion of pump materials by fly ash slurry using taguchi's orthogonal array. *Tribol. - Finnish J. Tribol.* 38, 31–38 (2021). https://doi.org/10.30678/fjt.97530.

51. Singh, J., Kumar, M., Kumar, S., Mohapatra, S.K.: Properties of glass-fiber hybrid composites: A review. *Polym. Plast. Technol. Eng.* 56, 455–469 (2017). https://doi.org/10.10 80/03602559.2016.1233271.
52. Singh, J., Mohapatra, S.K., Kumar, S.: Performance analysis of pump materials employed in bottom ash slurry erosion conditions. *J. Tribol.* 30, 73–89 (2021).
53. Singh, J.: A review on mechanisms and testing of wear in slurry pumps, pipeline circuits and hydraulic turbines. *J. Tribol.* 143, 1–83 (2021). https://doi.org/10.1115/1.4050977.
54. Singh, J., Singh, J.P.: Numerical analysis on solid particle erosion in elbow of a slurry conveying circuit. *J. Pipeline Syst. Eng. Pract.* 12, 04020070 (2021). https://doi.org/10.1061/(asce)ps.1949-1204.0000518.
55. Singh, J., Kumar, S., Singh, J.P., Kumar, P., Mohapatra, S.K.: CFD modeling of erosion wear in pipe bend for the flow of bottom ash suspension. *Part. Sci. Technol.* 37, 275–285 (2019). https://doi.org/10.1080/02726351.2017.1364816.
56. Singh, J., Kumar, S., Mohapatra, S.: Study on role of particle shape in erosion wear of austenitic steel using image processing analysis technique. *Proc. Inst. Mech. Eng. Part J J. Eng. Tribol.* 233, 712–725 (2019). https://doi.org/10.1177/1350650118794698.
57. Singh, M., Vasudev, H., Kumar, R.: Microstructural characterization of BN thin films using RF magnetron sputtering method. *Mater. Today Proc.* 26, 2277–2282 (2020). https://doi.org/10.1016/j.matpr.2020.02.493.
58. Vasudev, H., Singh, G., Bansal, A., Vardhan, S., Thakur, L.: Microwave heating and its applications in surface engineering: A review. *Mater. Res. Express.* 6, 102001 (2019). https://doi.org/10.1088/2053-1591/ab3674.
59. Prashar, G., Vasudev, H.: Structure–property correlation of plasma-sprayed Inconel625-Al$_2$O$_3$ bimodal composite coatings for high-temperature oxidation protection. *J. Therm. Spray Technol.* 31, 2385–2408 (2022). https://doi.org/10.1007/s11666-022-01466-1.
60. Vasudev, H., Thakur, L., Singh, H., Bansal, A.: Mechanical and microstructural behaviour of wear resistant coatings on cast iron lathe machine beds and slides. *Met. Mater.* 56, 55–63 (2018). https://doi.org/10.4149/km_2018_1_55.
61. Vasudev, H., Thakur, L., Singh, H., Bansal, A.: An investigation on oxidation behaviour of high velocity oxy-fuel sprayed Inconel718-Al$_2$O$_3$ composite coatings. *Surf. Coatings Technol.* 393, 125770 (2020). https://doi.org/10.1016/j.surfcoat.2020.125770.
62. Vasudev, H., Thakur, L., Bansal, A., Singh, H., Zafar, S.: High temperature oxidation and erosion behaviour of HVOF sprayed bi-layer Alloy-718/NiCrAlY coating. *Surf. Coatings Technol.* 362, 366–380 (2019). https://doi.org/10.1016/j.surfcoat.2019.02.012.
63. Prashar, G., Vasudev, H.: Structure-property correlation and high-temperature erosion performance of Inconel625-Al$_2$O$_3$ plasma-sprayed bimodal composite coatings. *Surf. Coatings Technol.* 439, 128450 (2022). https://doi.org/10.1016/j.surfcoat.2022.128450.
64. Vasudev, H.: Wear characteristics of Ni-WC powder deposited by using a microwave route on Mild steel. *Int. J. Surf. Eng. Interdiscip. Mater. Sci.* 8, 44–54 (2020). https://doi.org/10.4018/IJSEIMS.2020010104.
65. Vasudev, H., Thakur, L., Singh, H., Bansal, A.: Effect of addition of Al$_2$O$_3$ on the high-temperature solid particle erosion behaviour of HVOF sprayed Inconel-718 coatings. *Mater. Today Commun.* 30, 103017 (2022). https://doi.org/10.1016/j.mtcomm.2021.103017.
66. Bansal, A., Vasudev, H., Sharma, A.K., Kumar, P.: Investigation on the effect of post weld heat treatment on microwave joining of the Alloy-718 weldment. *Mater. Res. Express.* 6, 086554 (2019). https://doi.org/10.1088/2053-1591/ab1d9a.
67. Singh, M., Vasudev, H., Kumar, R.: Corrosion and tribological behaviour of BN thin films deposited using magnetron sputtering. *Int. J. Surf. Eng. Interdiscip. Mater. Sci.* 9, 24–39 (2021). https://doi.org/10.4018/IJSEIMS.2021070102.
68. Singh, M., Vasudev, H., Singh, M.: Surface protection of SS-316L with boron nitride based thin films using radio frequency magnetron sputtering technique. *J. Electrochem. Sci. Eng.* 00, 1–13 (2022). https://doi.org/10.5599/jese.1247.

69. Vasudev, H., Thakur, L., Singh, H., Bansal, A.: A study on processing and hot corrosion behaviour of HVOF sprayed Inconel718-nano Al₂O₃ coatings. *Mater. Today Commun.* 25, 101626 (2020). https://doi.org/10.1016/j.mtcomm.2020.101626.

70. Vasudev, H., Prashar, G., Thakur, L., Bansal, A.: Electrochemical corrosion behavior and microstructural characterization of HVOF sprayed inconel718-Al₂O₃ composite coatings. *Surf. Rev. Lett.* 29, 2250017 (2022). https://doi.org/10.1142/S0218625X22500172.

71. Prashar, G., Vasudev, H., Thakur, L.: High-temperature oxidation and erosion resistance of ni-based thermally-sprayed coatings used in power generation machinery: A review. *Surf. Rev. Lett.* 29, 2230003 (2022). https://doi.org/10.1142/S0218625X22300039.

72. Singh, P., Vasudev, H., Bansal, A.: Effect of post-heat treatment on the microstructural, mechanical, and bioactivity behavior of the microwave-assisted alumina-reinforced hydroxyapatite cladding. *Proc. Inst. Mech. Eng. Part E J. Process Mech. Eng.* 095440892211161 (2022). https://doi.org/10.1177/09544089221116168.

73. Singh, P., Bansal, A., Vasudev, H., Singh, P.: In situ surface modification of stainless steel with hydroxyapatite using microwave heating. *Surf. Topogr. Metrol. Prop.* 9, 035053 (2021). https://doi.org/10.1088/2051-672X/ac28a9.

74. Vasudev, H., Prashar, G., Thakur, L., Bansal, A.: Microstructural characterization and electrochemical corrosion behaviour of HVOF sprayed Alloy718-nanoAl₂O₃ composite coatings. *Surf. Topogr. Metrol. Prop.* 9, 035003 (2021). https://doi.org/10.1088/2051-672X/ac1044.

75. Singh, G., Vasudev, H., Bansal, A., Vardhan, S.: Influence of heat treatment on the microstructure and corrosion properties of the Inconel-625 clad deposited by microwave heating. *Surf. Topogr. Metrol. Prop.* 9, 025019 (2021). https://doi.org/10.1088/2051-672X/abfc61.

76. Prashar, G., Vasudev, H.: A review on the influence of process parameters and heat treatment on the corrosion performance of ni-based thermal spray coatings. *Surf. Rev. Lett.* 29, 1–18 (2022). https://doi.org/10.1142/S0218625X22300015.

77. Singh, J., Gill, H.S., Vasudev, H.: Computational fluid dynamics analysis on effect of particulate properties on erosive degradation of pipe bends. *Int. J. Interact. Des. Manuf.* (2022). https://doi.org/10.1007/s12008-022-01094-7.

78. Satyavathi Yedida, V.V., Vasudev, H.: A review on the development of thermal barrier coatings by using thermal spray techniques. *Mater. Today Proc.* 50, 1458–1464 (2022). https://doi.org/10.1016/j.matpr.2021.09.018.

79. K, S., Vasudev, H.: Microsrtructural and mechanical characterization of HVOF-sprayed Ni-based alloy coating. *Int. J. Surf. Eng. Interdiscip. Mater. Sci.* 10, 1–9 (2022). https://doi.org/10.4018/IJSEIMS.298705.

80. Mehta, A., Vasudev, H., Singh, S.: Recent developments in the designing of deposition of thermal barrier coatings – A review. *Mater. Today Proc.* 26, 1336–1342 (2020). https://doi.org/10.1016/j.matpr.2020.02.271.

81. Prashar, G., Vasudev, H.: Hot corrosion behavior of super alloys. *Mater. Today Proc.* 26, 1131–1135 (2020). https://doi.org/10.1016/j.matpr.2020.02.226.

82. Singh, J., Vasudev, H., Singh, S.: Performance of different coating materials against high temperature oxidation in boiler tubes – A review. *Mater. Today Proc.* 26, 972–978 (2020). https://doi.org/10.1016/j.matpr.2020.01.156.

83. Prashar, G., Vasudev, H.: Surface topology analysis of plasma sprayed Inconel625-Al₂O₃ composite coating. *Mater. Today Proc.* 50, 607–611 (2022). https://doi.org/10.1016/j.matpr.2021.03.090.

84. Zheng Yang, K., Pramanik, A., Basak, A.K., Dong, Y., Prakash, C., Shankar, S., Dixit, S., Kumar, K., Ivanovich Vatin, N.: Application of coolants during tool-based machining – A review. *Ain Shams Eng. J.* (2022). https://doi.org/10.1016/j.asej.2022.101830.

85. Pramanik, A., Basak, A.K., Littlefair, G., Debnath, S., Prakash, C., Singh, M.A., Marla, D., Singh, R.K.: Methods and variables in Electrical discharge machining of titanium alloy – A review. *Heliyon.* 6, e05554 (2020). https://doi.org/10.1016/j.heliyon.2020.e05554.

86. Prakash, C., Kansal, H.K., Pabla, B., Puri, S., Aggarwal, A.: Electric discharge machining – A potential choice for surface modification of metallic implants for orthopedic applications: A review. *Proc. Inst. Mech. Eng. Part B J. Eng. Manuf.* 230, 331–353 (2016). https://doi.org/10.1177/0954405415579113.

87. Prakash, C., Kansal, H.K., Pabla, B.S., Puri, S.: Processing and characterization of novel biomimetic nanoporous bioceramic surface on β-Ti implant by powder mixed electric discharge machining. *J. Mater. Eng. Perform.* 24, 3622–3633 (2015). https://doi.org/10.1007/s11665-015-1619-6.

88. Pradhan, S., Singh, S., Prakash, C., Królczyk, G., Pramanik, A., Pruncu, C.I.: Investigation of machining characteristics of hard-to machine Ti-6Al-4V-ELI alloy for biomedical applications. *J. Mater. Res. Technol.* 8, 4849–4862 (2019). https://doi.org/10.1016/j.jmrt.2019.08.033.

89. Innocentini, M.D.M., Botti, R.F., Bassi, P.M., Paschoalato, C.F.P.R., Flumignan, D.L., Franchin, G., Colombo, P.: Lattice-shaped geopolymer catalyst for biodiesel synthesis fabricated by additive manufacturing. *Ceram. Int.* 45, 1443–1446 (2019). https://doi.org/10.1016/j.ceramint.2018.09.239.

90. Kumar, P., Kumar, S., Singh, J.: Effect of natural surfactant on the rheological characteristics of heavy crude oil. *Mater. Today Proc.* 5, 23881–23887 (2018). https://doi.org/10.1016/j.matpr.2018.10.180.

91. Kumar, P., Singh, J., Singh, S.: Neural network supported flow characteristics analysis of heavy sour crude oil emulsified by ecofriendly bio-surfactant utilized as a replacement of sweet crude oil. *Chem. Eng. J. Adv.* 11, 100342 (2022). https://doi.org/10.1016/j.ceja.2022.100342.

92. Kumar, P., Kumar, S., Singh, J.: Rheological and computational analysis of crude oil transportation. *Int. J. Mech. Aerospace, Ind. Mechatron. Manuf. Eng.* 11, 429–432 (2017).

93. Kumar, P., Singh, J.: Computational study on effect of Mahua natural surfactant on the flow properties of heavy crude oil in a 90° bend. *Mater. Today Proc.* 43, 682–688 (2021). https://doi.org/10.1016/j.matpr.2020.12.612.

94. Yoo, D.-J.: Recent trends and challenges in computer-aided design of additive manufacturing-based biomimetic scaffolds and bioartificial organs. *Int. J. Precis. Eng. Manuf.* 15, 2205–2217 (2014). https://doi.org/10.1007/s12541-014-0583-7.

95. Bastas, A.: Sustainable manufacturing technologies: A systematic review of latest trends and themes. *Sustainability.* 13, 4271 (2021). https://doi.org/10.3390/su13084271.

96. Chu, W.-S., Kim, M.-S., Jang, K.-H., Song, J.-H., Rodrigue, H., Chun, D.-M., Cho, Y.T., Ko, S.H., Cho, K.-J., Cha, S.W., Min, S., Jeong, S.H., Jeong, H., Lee, C.-M., Chu, C.N., Ahn, S.-H.: From design for manufacturing (DFM) to manufacturing for design (MFD) via hybrid manufacturing and smart factory: A review and perspective of paradigm shift. *Int. J. Precis. Eng. Manuf. Technol.* 3, 209–222 (2016). https://doi.org/10.1007/s40684-016-0028-0.

97. Gibson, I., Rosen, D.W., Stucker, B.: Development of additive manufacturing technology. In: Gibson, I., Rosen, D.W., and Stucker, B. (eds.) *Additive Manufacturing Technologies.* pp. 19–42. Springer, Berlin, Heidelberg (2010).

98. Ahn, D.-G.: Applications of laser assisted metal rapid tooling process to manufacture of molding & forming tools — state of the art. *Int. J. Precis. Eng. Manuf.* 12, 925–938 (2011). https://doi.org/10.1007/s12541-011-0125-5.

99. Sun, S., Brandt, M., Easton, M.: Powder bed fusion processes. In: *Laser Additive Manufacturing.* pp. 55–77. Elsevier (2017). https://doi.org/10.1016/B978-0-08-100433-3.00002-6.

100. Cristofolini, I., Pilla, M., Rao, A., Libardi, S., Molinari, A.: Dimensional and geometrical precision of powder metallurgy parts sintered and sinterhardened at high temperature. *Int. J. Precis. Eng. Manuf.* 14, 1735–1742 (2013). https://doi.org/10.1007/s12541-013-0233-5.

101. Lee, H.-J., Song, J.-G., Ahn, D.-G.: Investigation into the influence of feeding parameters on the formation of the fed-powder layer in a powder bed fusion (PBF) system. *Int. J. Precis. Eng. Manuf.* 18, 613–621 (2017). https://doi.org/10.1007/s12541-017-0073-9.
102. Mahamood, R.M., Akinlabi, E.T.: Laser additive manufacturing. In: Esther, A., Mahamood, T., Akinlabi, R., and Akinwale, S. (eds.) *Advanced Manufacturing Techniques Using Laser Material Processing.* pp. 1–23. IGI Global, Hershey, PA (2016). https://doi.org/10.4018/978-1-5225-0329-3.ch001.
103. Wohlers, T.: Wohlers Report 2013: Additive Manufacturing and 3D Printing, State of the Industry–Annual Worldwide Progress Report., Fort Collins, CO (2013).
104. EY's Global 3D Printing Report: Executive Summary- How will 3D Printing Make your Company the Strongest Link in the Value Chain? (2016).
105. Wohlers, T.: Wohlers Report 2014: 3D Printing and Additive Manufacturing State of the Industry., Fort Collins, CO (2014).
106. Belforte, D.: 2015 Industrial Laser Market Outperforms Global Manufacturing Instability. (2015).
107. Markets: 3D Printing Metal Market by Form (Powder and Filament), by Type (Titanium, Nickel, Stainless Steel, Aluminum, Others), by Application (Aerospace & Defense, Automotive, Medical & Dental, Others), and by Region - Global Forecast to 2020. (2016).
108. Huang, C., Geng, J., Luo, T., Han, J., Wang, Q., Liang, R., Fan, S., Jiang, S.: Rare earth doped optical fibers with multi-section core. *iScience.* 22, 423–429 (2019). https://doi.org/10.1016/j.isci.2019.11.017.
109. Patel, C.K.N.: Continuous-wave laser action on vibrational-rotational transitions of CO_2. *Phys. Rev.* 136, A1187–A1193 (1964). https://doi.org/10.1103/PhysRev.136.A1187.
110. Witteman, W.J.: Continuous discharge lasers. In: Witteman, W.J. (ed.) *The CO_2 Laser.* pp. 81–126. Springer, Berlin, Heidelberg (1987). https://doi.org/10.1007/978-3-540-47744-0_4.
111. Bass, M.: *Laser Materials Processing.* Elsevier Ltd Amsterdam, Netherlands (2012).
112. Majumdar, J.D., Manna, I.: *Laser-Assisted Fabrication of Materials.* Springer Science & Business Media, Berlin, Germany (2012).
113. Witteman, W.J.: Introduction. In: Witteman, W.J. (ed.) *The CO_2 Laser.* pp. 1–7. Springer, Berlin, Heidelberg (1987). https://doi.org/10.1007/978-3-540-47744-0_1.
114. Singh, J., Singh, C.: Computational analysis of convective heat transfer across a vertical tube. *FME Trans.* 49, 932–940 (2021). https://doi.org/10.5937/fme2104932S.
115. Singh, J., Singh, C.: Study of buoyant force acting on different fluids moving over a horizontal plate due to forced convection. *IOP Conf. Ser. Mater. Sci. Eng.* 710, 012044 (2019). https://doi.org/10.1088/1757-899X/710/1/012044.
116. Singh, J., Singh, C.: Numerical analysis of heat dissipation from a heated vertical cylinder by natural convection. *Proc. Inst. Mech. Eng. Part E J. Process Mech. Eng.* 231, 405–413 (2017). https://doi.org/10.1177/0954408915600109.
117. Tredicce, J.R., Quel, E.J., Ghazzawi, A.M., Green, C., Pernigo, M.A., Narducci, L.M., Lugiato, L.A.: Spatial and temporal instabilities in a CO_2 laser. *Phys. Rev. Lett.* 62, 1274–1277 (1989). https://doi.org/10.1103/PhysRevLett.62.1274.
118. Nighan, W.L., Wiegand, W.J., Haas, R.A.: Ionization instability in CO_2 laser discharges. *Appl. Phys. Lett.* 22, 579–582 (1973). https://doi.org/10.1063/1.1654515.
119. Digonnet, M., Gaeta, C., Shaw, H.: 1.064- and 1.32-μm Nd:YAG single crystal fiber lasers. *J. Light. Technol.* 4, 454–460 (1986). https://doi.org/10.1109/JLT.1986.1074730.
120. Farças, I.I.: Development of laser material processing in Romania. In: Martellucci, S., Chester, A.N., and Scheggi, A.M. (eds.) *Laser Applications for Mechanical Industry.* pp. 283–290. Springer Science & Business Media, Berlin, Germany (1993).
121. Geusic, J.E., Marcos, H.M., Van Uitert, L.G.: Laser oscillations in nd-doped yttrium aluminum, yttrium gallium and gadolinium garnets. *Appl. Phys. Lett.* 4, 182–184 (1964). https://doi.org/10.1063/1.1753928.

122. Maina, M.R., Okamoto, Y., Hamada, K., Okada, A., Nakashiba, S., Nishi, N.: Effects of superposition of 532 nm and 1064 nm wavelengths in copper micro-welding by pulsed Nd:YAG laser. *J. Mater. Process. Technol.* 299, 117388 (2022). https://doi.org/10.1016/j.jmatprotec.2021.117388.

123. Weber, R., Neuenschwander, B., Weber, H.P.: Thermal effects in solid-state laser materials. *Opt. Mater.* 11, 245–254 (1999). https://doi.org/10.1016/S0925-3467(98)00047-0.

124. Berger, J., Hoffman, N.J., Smith, J.J., Welch, D.F., Streifer, W., Radecki, D., Scifres, D.R.: Fiber-bundle coupled, diode end-pumped Nd:YAG laser. *Opt. Lett.* 13, 306 (1988). https://doi.org/10.1364/OL.13.000306.

125. Zhou, B., Kane, T.J., Dixon, G.J., Byer, R.L.: Efficient, frequency-stable laser-diode-pumped Nd:YAG laser. *Opt. Lett.* 10, 62 (1985). https://doi.org/10.1364/OL.10.000062.

126. Hügel, H.: New solid-state lasers and their application potentials. *Opt. Lasers Eng.* 34, 213–229 (2000). https://doi.org/10.1016/S0143-8166(00)00065-8.

127. Kruth, J.P., Kumar, S., Van Vaerenbergh, J.: Study of laser-sinterability of ferro-based powders. *Rapid Prototyp. J.* 11, 287–292 (2005). https://doi.org/10.1108/13552540510623594.

128. Mumtaz, K., Hopkinson, N.: Selective laser melting of Inconel 625 using pulse shaping. *Rapid Prototyp. J.* 16, 248–257 (2010). https://doi.org/10.1108/13552541011049261.

129. Kobryn, P.A., Semiatin, S.L.: The laser additive manufacture of Ti-6Al-4V. *JOM.* 53, 40–42 (2001). https://doi.org/10.1007/s11837-001-0068-x.

130. Liao, H., Shie, J.: Optimization on selective laser sintering of metallic powder via design of experiments method. *Rapid Prototyp. J.* 13, 156–162 (2007). https://doi.org/10.1108/13552540710750906.

131. Balla, V.K., Bose, S., Bandyopadhyay, A.: Processing of bulk alumina ceramics using laser engineered net shaping. *Int. J. Appl. Ceram. Technol.* 5, 234–242 (2008). https://doi.org/10.1111/j.1744-7402.2008.02202.x.

133. Garg, A., Lam, J.S.L., Savalani, M.M.: Laser power based surface characteristics models for 3-D printing process. *J. Intell. Manuf.* 29, 1191–1202 (2018). https://doi.org/10.1007/s10845-015-1167-9.

133. Minassian, A., Thompson, B., Damzen, M.J.: Ultrahigh-efficiency TEM 00 diode-side-pumped Nd:YVO 4 laser. *Appl. Phys. B Lasers Opt.* 76, 341–343 (2003). https://doi.org/10.1007/s00340-003-1095-9.

134. Délen, X., Balembois, F., Musset, O., Georges, P.: Characteristics of laser operation at 1064 nm in Nd:YVO_4 under diode pumping at 808 and 914 nm. *J. Opt. Soc. Am. B.* 28, 52 (2011). https://doi.org/10.1364/JOSAB.28.000052.

135. Huang, B.W., Chen, M.Y.: Evaluation on some properties of SL7560 type photosensitive resin and its fabricated parts. *Appl. Mech. Mater.* 117–119, 1164–1167 (2011). https://doi.org/10.4028/www.scientific.net/AMM.117-119.1164.

136. Huang, B.W., Weng, Z.X., Sun, W.: Study on the properties of DSM SOMOS 11120 type photosensitive resin for stereolithography materials. *Adv. Mater. Res.* 233–235, 194–197 (2011). https://doi.org/10.4028/www.scientific.net/AMR.233-235.194.

137. Dutta, N.K.: *Fiber Amplifiers and Fiber Lasers*. World Scientific Publishing Co Pte Ltd, Singapore (2014).

138. Muniz-Cánovas, P., Barmenkov, Y.O., Kir'yanov, A. V., Cruz, J.L., Andrés, M. V.: Ytterbium-doped fiber laser as pulsed source of narrowband amplified spontaneous emission. *Sci. Rep.* 9, 13073 (2019). https://doi.org/10.1038/s41598-019-49695-9.

139. Verhaeghe, G., Hilton, P.: Battles of the Sources-Using a High- Power Yb-Fibre Laser for Welding Steel and Aluminium. In: *Proceeding of the 3rd International WLT Conference in Manufacturing.* pp. 33–38 (2005).

140. Limpert, J., Schreiber, T., Nolte, S., Zellmer, H., Tunnermann, T., Iliew, R., Lederer, F., Broeng, J., Vienne, G., Petersson, A., Jakobsen, C.: High-power air-clad large-mode-area photonic crystal fiber laser. *Opt. Express.* 11, 818 (2003). https://doi.org/10.1364/OE.11.000818.

141. Kong, F., Gu, G., Hawkins, T.W., Parsons, J., Jones, M., Dunn, C., Kalichevsky-Dong, M.T., Wei, K., Samson, B., Dong, L.: Flat-top mode from a 50 μm-core Yb-doped leakage channel fiber. *Opt. Express.* 21, 32371 (2013). https://doi.org/10.1364/OE.21.032371.
142. Sezerman, O., Best, G.: Accurate alignment preserves polarization. *Laser Focus World.* 33, S27–S30 (1997).
143. Basting, D., Pippert, K.D., Stamm, U.: History and future prospects of excimer lasers. *Proc. SPIE.* 4426, 25–26 (2002).
144. Mann, K.R., Eva, E.: Characterizing the absorption and aging behavior of DUV optical material by high-resolution excimer laser calorimetry. *Proc. SPIE.* 3334, 1055–1061 (1998).
145. Morozov, N. V.: Laser-induced damage in optical materials under uv excimer laser radiation. *Proc. SPIE.* 2428, 153–169 (1995).
146. Atezhev, V. V., Vartapetov, S.K., Zhukov, A.N., Kurzanov, M.A., Obidin, A.Z.: Excimer laser with highly coherent radiation. *Quantum Electron.* 33, 689–694 (2003). https://doi.org/10.1070/QE2003v033n08ABEH002479.
147. Toenshoff, H.K., Ostendorf, A., Koerber, K., Meyer, K.: Comparison of machining strategies for ceramics using frequency-converted Nd: YAG and excimer lasers. *Proc. SPIE.* 4426, 408–411 (2002).
148. Tuck, C.: priv comm., (2010).
149. Beaman, J.J., Barlow, J.W., Bourell, D.L., Crawford, R.H., Marcus, H.L., McAlea, K.P.: *Solid Freeform Fabrication: A New Direction In Manufacturing.* Kluwer Academic Press, Boston, MA (1997).
150. Zhao, H.-B., Pflanz, K., Gu, J.-H., Li, A.-W., Stroh, N., Brunner, H., Xiong, G.-X.: Preparation of palladium composite membranes by modified electroless plating procedure. *J. Memb. Sci.* 142, 147–157 (1998). https://doi.org/10.1016/S0376-7388(97)00287-1.
151. Palaniappa, M., Babu, G.V., Balasubramanian, K.: Electroless nickel–phosphorus plating on graphite powder. *Mater. Sci. Eng. A.* 471, 165–168 (2007). https://doi.org/10.1016/j.msea.2007.03.004.
152. Chen, L.F., Luo, X.T., Wu, Q.L.: Electrochemical preparation of fiber reinforced metallic matrix composites. *J. Mater. Sci. Lett.* 23, 379–381 (2003).
153. Dück, J., Niebling, F., Neeße, T., Otto, A.: Infiltration as post-processing of laser sintered metal parts. *Powder Technol.* 145, 62–68 (2004). https://doi.org/10.1016/j.powtec.2004.05.006.
154. Xu, C., Li, M., Zhang, X., Tu, K.-N., Xie, Y.: Theoretical studies of displacement deposition of nickel into porous silicon with ultrahigh aspect ratio. *Electrochim. Acta.* 52, 3901–3909 (2007). https://doi.org/10.1016/j.electacta.2006.11.007.
155. Holmström, J., Gutowski, T.: Additive manufacturing in operations and supply chain management: No sustainability benefit or virtuous knock-on opportunities? *J. Ind. Ecol.* 21, S21–S24 (2017). https://doi.org/10.1111/jiec.12580.
156. Hunt, E.J., Zhang, C., Anzalone, N., Pearce, J.M.: Polymer recycling codes for distributed manufacturing with 3-D printers. *Resour. Conserv. Recycl.* 97, 24–30 (2015). https://doi.org/10.1016/j.resconrec.2015.02.004.
157. Hao, L., Raymond, D., Strano, G., Dadbakhsh, S.: Enhancing the sustainability of additive manufacturing. In: *IET Conference Publications.* p. 390 (2010).
158. Mani, M., Lyons, K.W., Gupta, S.K.: Sustainability characterization for additive manufacturing. *J. Res. Natl. Inst. Stand. Technol.* 119, 419 (2014). https://doi.org/10.6028/jres.119.016.
159. Ford, S., Despeisse, M.: Additive manufacturing and sustainability: An exploratory study of the advantages and challenges. *J. Clean. Prod.* 137, 1573–1587 (2016). https://doi.org/10.1016/j.jclepro.2016.04.150.
160. Despeisse, M., Baumers, M., Brown, P., Charnley, F., Ford, S.J., Garmulewicz, A., Knowles, S., Minshall, T.H.W., Mortara, L., Reed-Tsochas, F.P., Rowley, J.: Unlocking value for a circular economy through 3D printing: A research agenda. *Technol. Forecast. Soc. Change.* 115, 75–84 (2017). https://doi.org/10.1016/j.techfore.2016.09.021.

Index

For Product Safety Concerns and Information please contact our EU
representative GPSR@taylorandfrancis.com
Taylor & Francis Verlag GmbH, Kaufingerstraße 24, 80331 München, Germany